BABY-MAKING

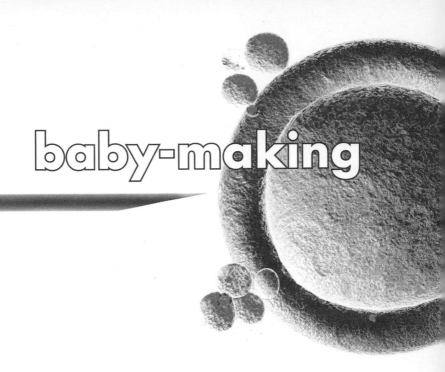

baby-making

BART FAUSER
& PAUL DEVROEY
with
SIMON BROWN

OXFORD
UNIVERSITY PRESS

OXFORD

UNIVERSITY PRESS

Great Clarendon Street, Oxford OX2 6DP

Oxford University Press is a department of the University of Oxford.
It furthers the University's objective of excellence in research, scholarship,
and education by publishing worldwide in

Oxford New York

Auckland Cape Town Dar es Salaam Hong Kong Karachi
Kuala Lumpur Madrid Melbourne Mexico City Nairobi
New Delhi Shanghai Taipei Toronto

With offices in

Argentina Austria Brazil Chile Czech Republic France Greece
Guatemala Hungary Italy Japan Poland Portugal Singapore
South Korea Switzerland Thailand Turkey Ukraine Vietnam

Oxford is a registered trade mark of Oxford University Press
in the UK and in certain other countries

Published in the United States
by Oxford University Press Inc., New York

British Library Cataloguing in Publication Data

Data available

Library of Congress Cataloging in Publication Data

Data available

Typeset by SPI Publisher Services, Pondicherry, India
Printed in Great Britain
on acid-free paper by
Clays Ltd, St Ives plc

ISBN 978-0-19-959731-4

1 3 5 7 9 10 8 6 4 2

CONTENTS

PREFACE

The treatment of infertility by assisted reproduction is a booming industry. The latest data—based on activity figures from the year 2007—suggest that around 1.5 million cycles of IVF and ICSI are performed each year, with numbers growing by at least 5 per cent annually. Within the next year or two, the global total of 'babies' conceived by IVF (many of whom now have children of their own) will reach 5 million, enough to populate a small country.

For the many millions of couples who are parents to these children infertility and its treatment are an intensely private matter, generating the greatest depths of frustration in its disappointments, and the heights of joy when treatment is successful. But today infertility and its many treatments are not just a personal matter. IVF and ICSI are performed within a social, political and economic context which have an enormous impact—direct or indirect—on how the treatments are applied, how safe they are, how much they cost, to what extent they are accessible, and who is likely to benefit the most. Some of those with most to gain, such as those at risk of passing on a genetic disease, may even be perfectly fertile.

So this is not a text book, nor a self-help manual for those confronting their own infertility. It is a statement of how we see assisted reproduction today, with its own agenda of low-risk and low-cost treatment. The picture we draw of IVF and the many treatments based upon this classic procedure is derived largely from our own experience and the scientific progress charted in the medical journals. Thus, our two separate clinics and the research we have done there frequently appear in the following pages—Paul Devroey's at the hospital of the Vrije Universiteit Brussel (VUB, Free University of Brussels) in Belgium, and Bart Fauser's at the University Medical Center at Utrecht, the Netherlands. References in the text to 'we' and 'our', particularly with respect to research, are often rhetorical and invariably imply the involvement of one or the other (usually clarified by 'Brussels' or 'Utrecht').

The content of the book is ours, developed and redeveloped in many lengthy sessions over the two years of 2009 and 2010. References to time, such as 'recent' or 'latest', should be understood within that context. While the language of assisted reproduction is English, it is not our mother tongue, and we are grateful to the medical writer Simon Brown for assisting us with the text and making linguistic sense of our ideas and observations.

However, as this book makes clear, the treatment of infertility is an international discipline applied in a small world of specialists and clinics. Progress in the field has been and remains rapid and exciting, although the days of major breakthroughs seem over, at least for the time being. So our perspective on this world of IVF is a broad international one, viewed from our vantage points at two of Europe's leading fertility centres.

What we see is not always what the public and policymakers see, which is why we wrote the book. Thus, what is here is not just an account of the technologies which constitute fertility treatment today, but also of that treatment within a context of advancing maternal age, falling fertility rates and continuing high rates of multiple pregnancy. In our view—and as made clear in this book—the ultimate objective of fertility treatment should be a healthy singleton live birth delivered to term and conceived without risk and at reasonable cost. The means of doing that are now within our grasp.

Bart Fauser, Paul Devroey, July 2011

LIST OF FIGURES

How to Design a Baby

The introduction of the contraceptive pill in the early 1960s was eventually hailed as a revolution in reproduction. More than any other contraceptive in history, the pill would take pregnancy out of sex, and remove the risk from the pleasure. For many women on the pill sex would now be about recreation, not procreation.

In its own way the first successful live birth following in vitro fertilisation (IVF) was no less a reproductive revolution. For what the birth of Louise Brown demonstrated, as she emerged into the world on a summer's night in 1978, was that procreation now need have little to do with recreation. IVF, like an immaculate conception, would take the sex out of pregnancy. Thus, while the pill allowed couples to have sex without children, IVF would allow them to have children without sex. And whether children are conceived with or without sexual intercourse holds the key to how babies can be designed.

Couples who conceive naturally while trying to get pregnant have little control over the outcome. Those who conceive while trying to avoid pregnancy have even less. All studies show that humans are inefficient breeders, and that even a healthy couple has no more than a one-in-four chance of conception in a single cycle of unprotected intercourse. For the subfertile—as many as one in six of the population—these odds are much less, and in some cases are as low as zero. And even in the event of a spontaneous conception there is little the couple can do about their baby's gender, its eye colour or its predisposition to illness. Such characteristics are left to natural, not parental, selection in a biological lottery which owes more to the laws of genetics than to the wishes of the couple.

In reality, the only control which a prospective parent has over a spontaneous pregnancy is partner selection. So a health check on a prospective partner/parent may be reason enough to say yes…or no. Already, companies in Europe and the US are offering individuals a 'complete scan' of their genetic risk profile as a guide to their future health. The tests are offered as an early warning scheme of personal risk, so that lifestyle changes or early treatments can be applied. But the same technology, made possible when mapping of the human genome was completed in 2003, might also forecast with some accuracy that a prospective husband is at high risk of early colon cancer or cystic fibrosis. Prenuptial agreements of the not too distant future might therefore require a comprehensive genomic profile of each partner, as well as the now familiar contingency plans over financial assets. All it would take is a saliva sample, and from that droplet an individual's DNA will hold the key to a multitude of gene variants (known as single nucleotide polymorphisms, or SNPs), some of which have now been associated with a susceptibility to disease—some serious—in later life. New technologies have made it possible to detect these gene variants in a human genome by comparing an individual's SNP pattern with that of a healthy reference group. Thus, a particular SNP pattern might be indicative of an individual's risk of a particular disease; if known publicly, such information might deter an insurer from writing a policy, or a man from finding a partner and becoming a father.

Choice of partner, of course, need not be so draconian, nor so scientific. Some biosocial studies have found a tendency for 'oedipal imprinting' by which individuals choose a partner with characteristics familiar in an opposite-sex parent. But most studies simply show that people tend to find partners of comparable physical attributes and backgrounds to themselves, without any obvious reason why. And what could be more non-selective than that?

The end result of this non-selection in human reproduction is that (almost) as many baby girls are born as baby boys, but not all of them are as healthy as their parents would wish. For example, of 708,711 babies born in England and Wales in 2008, 362,963 were boys and 345,748 girls, a gender ratio of 1.05 to 1.0. In the US in 2005 this gender ratio was exactly the same—reflecting a proportion of 51 per cent boys and 49 per cent girls. Nowhere is the Mendelian law of genetics better illustrated than in this relatively equal division of boys and girls in any naturally selected birth cohort—that each parent provides a gene pair from which (in the egg or sperm cell) one copy has separated in a random fashion to unite with a copy from the other. The laws of natural selection would thus dictate that, of the 46 chromosomes in a normal human cell, half are derived from the mother's egg and half are derived from the father's sperm. When the sperm enters the egg at fertilisation to produce a new organism, it has a full complement of 23 chromosome pairs, each one carrying the inherited genes, but how these chromosomes are sorted is a random selection derived from all possible combinations of maternal and paternal chromosomes.

Moreover, in this same British birth cohort of 700,000 babies as many as 3 per cent will have some kind of (major or minor) inherited abnormality, in which defective genes are passed down from one generation to the next or derived from a spontaneous mutation. Some of these inheritances will be evident as serious diseases caused by single gene defects (such as cystic fibrosis or certain types of muscular dystrophy), others evident in little more than a slight deformity of a toe. There will be cases too of congenital heart disease, and others where a certain susceptibility to illness will require time and environmental influence to express itself (as in diabetes or hypertension). In addition, around one in a thousand—some 700 of all babies born in Britain in 2008—will be born with a chromosomal abnormality known as Down's syndrome, a condition in which there are three copies of chromosome 21, instead of the usual pair. A 2009 study confirmed that the risk of a Down's syndrome baby increases with maternal age, with prevalence found to be 39 per 10,000 babies in older mothers (defined as over 35), and eight per 10,000 in younger.[1] However, the true prevalence of Down's syndrome is substantially higher than this, because many pregnancies affected by trisomy 21 are spontaneously miscarried or deliberately terminated. With more and more women postponing their first pregnancy, the prevalence of Down's syndrome is increasing, but its presence is still perceived as a random event derived from a numerical misarrangement of chromosomes.

By contrast, assisted reproduction removes some—but not all—of this serendipity from conception. In modern IVF, for example, the ovaries are stimulated with fertility drugs to produce several eggs. Most of them will fertilise when exposed to sperm in a test-tube, but only the most healthy-looking embryos are selected for transfer into the uterus. Similarly, in the technique of intracytoplasmic sperm injection (or ICSI) the best-looking sperm cells are also selected for fertilisation. In spontaneous conception as many as 20 million sperms will be deposited in the vagina with each ejaculation, one of which may swim through the cervix to meet and finally penetrate an ovulated egg and form an embryo. In ICSI, however, beneath the lens of the most powerful microscopes, just one carefully selected sperm cell in the hands of a skilled embryologist will be sufficient. And where an estimated fertilisation rate may be no better than 50 per cent in natural conception, in ICSI almost every egg injected with a sperm cell will fertilise and become an embryo, with fertilisation rates as high as 90 per cent recorded in some clinics.

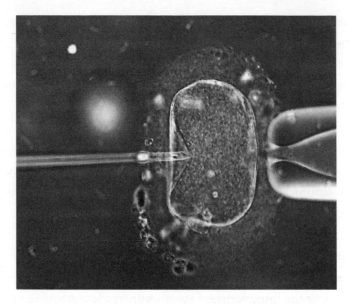

Figure 1 For the first time, ICSI made fatherhood possible for many men with sperm defects. Fertilisation is achieved by the injection of a single sperm cell into the cytoplasm of an oocyte. ICSI was developed at the VUB in Brussels in the early 1990s.

ICSI results from our centre at the hospital of the Dutch-speaking Free University of Brussels (VUB) show that fertilisation rate per injected oocyte is between 60 and 90 per cent depending on female age and sperm quality. Thus, today there are two basic techniques of assisted reproduction available for the treatment of infertile couples, the first the traditional IVF technique which is predominantly indicated for female infertility in the presence of normal, healthy spermatozoa. A good example here would be a woman with absent or damaged Fallopian tubes. In a case such as this, the IVF lab is in effect taking over the activity of the Fallopian tube and providing a substitute environment for fertilisation.

However, in cases of male factor infertility—usually because of a condition known as 'oligospermia', which implies too few or too many immotile or mis-shaped sperm cells—ICSI now provides a second and very successful fertilisation technique. ICSI was developed at the VUB and first described in a report to the medical journal *The Lancet* in 1992.[2] At the time, it was truly astonishing for us at the VUB to see that even in extremely poor cases, where only a few hundred spermatozoa were present, normal fertilisation rates could still be achieved.

And it was even more remarkable to discover that the final outcome—pregnancy and live birth—was related not to the quality of the impoverished sperm sample but to the age of the female partner. This is a finding which has dominated assisted reproduction since that time, that the most important prognostic factor for outcome, in terms of egg quality and the implantation potential of the embryo, is the age of the woman.

Since then, we have shown that even in azoospermic male patients, where no sperms of any description are present in the ejaculate, sperm cells suitable for ICSI can be retrieved surgically from the testis. We usually encounter two different conditions: either a 'mechanical' problem causing an obstruction, or a non-obstructive physiological problem. The former might be caused by a damaged epididymis (part of the tubal network from the testis in which sperm is stored) but even in these once insoluble cases we are now able to retrieve sperm surgically in a procedure known as testicular sperm extraction (TESE) and, after isolation, inject a single sperm cell into the female egg. The same procedure can be used in cases of non-obstructive azoospermia—for instance, in male patients with small testes and very few spermatozoa produced—although sperm cells are then more difficult to find and will be extracted in only 50 per cent of cases.

Before our first experiments in treating these difficult cases with TESE, it was our belief—in common with many reproductive scientists—that a sperm cell needed to pass the epididymis before it reached its full maturation. But when we took testicular biopsies in these cases of obstructive azoospermia, we were astonished to see the sperm cells moving under the microscope. These were the sperms we selected for injection with ICSI. We published our findings in 1996, and from then on it was clear that even men with an obstruction in the tubal network from the testis could now become fathers of their own biological children.

So historically, conventional IVF has been developed for those types of infertility whose cause lies with the female partner, and ICSI for male factor infertility. However, what we know today is that, whichever technique is used or indicated, the final chance of success is dominated by the age of the woman. The pregnancy rate in women under the age of 25 will be significantly higher than in women over the age of 40, with a pregnancy rate of around 50 per cent per cycle in the former declining to no more than 10 per cent in the latter.

Today, whatever the cause of infertility and whether treatment is by IVF or ICSI, successful programmes nearly always include ovarian stimulation to produce not one but several eggs. The idea is that these will produce several embryos from which the embryologist can select the ones which seem under the microscope to be the most likely to implant in the uterus.

However, it is often forgotten that the birth of Louise Brown, the world's first IVF baby, was derived not from one egg selected from several following ovarian stimulation, but from the single ovulatory follicle of a natural menstrual cycle. Back in the summer of 1977 Lesley Brown, the mother of Louise, had been referred to the gynaecologist Patrick Steptoe because of his skills with a new 'keyhole' instrument called a laparoscope. Laparoscopy, it was thought, might provide an accurate diagnosis of Lesley's infertility without the need for open surgery—and laparoscopy did indeed confirm 'grossly distorted tubal remnants' with occlusions and adhesions in her Fallopian tube.

It was Steptoe's skills in laparoscopy which had also attracted the attention of the Cambridge biologist Robert Edwards, who saw in this new instrument an efficient means of retrieving eggs from the ovaries of ovulating women. At last, thought Edwards, here was the promise of raw material for his numerous experiments in human fertilisation—and the way to overcome the paucity of live human eggs which had plagued researchers in reproduction for the previous 20 years. Laparoscopy, of which Steptoe was the British pioneer, allowed entry into the abdominal cavity through two small holes: one for the surgical instrument and the other for a camera which would display its images on a TV screen.

Louise Brown was not the world's first fertilisation in vitro (in 'glass', hence the 'test-tube' baby), nor the first IVF pregnancy, but she was the first live birth. The first report of a pregnancy following fertilisation in vitro had actually come from Melbourne, Australia, in 1973 but transfer of the embryo had been followed by an early miscarriage. Three years later, Edwards and Steptoe themselves had described the transfer of a well-developed embryo into the mother's uterus. Monitoring, however, showed that the foetus was growing in a reproductive tube as an ectopic pregnancy, which was terminated at 13 weeks. But that pregnancy, albeit ectopic, did confirm that embryos cultured in the laboratory could implant and grow to the advanced stages of cell division consistent with a healthy pregnancy.

The 1978 report of Steptoe and Edwards to the *Lancet* describing the birth of Louise Brown states that the pregnancy was established following 'the laparoscopic recovery of an oocyte', which was in fact the oocyte generated from the leading follicle of Lesley's natural cycle.[3] Thus, the world's first IVF baby was actually the result of a natural experiment in which the ovulating egg of a normal menstrual cycle was captured by laparoscopy and fertilised with a sperm sample from Lesley's husband. This first successful IVF, therefore, had circumvented the occlusions in Lesley's Fallopian tube and brought together in the laboratory what nature had always intended.

The achievement of Louise Brown's birth, hailed throughout the world as a landmark of monumental if controversial proportions, set a new—albeit short-lived—blueprint for the successful application of IVF: egg collection from a natural menstrual cycle. But that naturally ovulating oocyte, produced by the healthy ovary every 28 days, proved stubbornly resistant to future success in IVF, and it would take another two years before the next test-tube birth was reported (this time from

Australia). Even the US's first dedicated centre for infertility, which was opened in Norfolk, Virginia, in 1980, was founded upon the natural and not the stimulated cycle. It would take more than 40 unsuccessful attempts in a natural cycle (with an egg retrieved in only 19 patients) before the Norfolk group of Howard and Georgeanna Jones finally gave up on nature and turned to the hormonally stimulated cycle and multiple eggs as a more likely way to success. Norfolk finally achieved its first pregnancy in the thirteenth patient to receive ovarian stimulation from fertility drugs. Delivery by Caesarean section followed in December 1981, and with it the unequivocal realisation that stimulated cycles and multiple eggs would produce better results in IVF than the sole leading follicle of a natural cycle. The difference was, of course, that multiple eggs from a stimulated cycle not only allowed the transfer of multiple embryos, but also allowed embryo selection and a degree of control which the natural cycle never did.

Since the birth of Louise Brown there have been more than 4 million IVF babies born throughout the world, and the principle of their conception is the same today as it was more than 30 years ago. That principle is based on the availability of gametes (eggs and sperm) in the laboratory for fertilisation, the culture and assessment of embryos, and their transfer and implantation in the uterus as a pregnancy.

Fertility drugs—known as gonadotrophins—to increase the number of female gametes were available long before Louise Brown's birth from natural cycle IVF. Indeed, the terminated ectopic pregnancy reported by Edwards and Steptoe in 1976 was the last in a series of eight patients all given gonadotrophins to stimulate the ovaries, and another hormone—human chorionic gonadotrophin (hCG)—to trigger final egg maturation. But there were no successful deliveries with a hormonal approach until 1981, when the Americans in Norfolk and Australians in Melbourne began reporting success after success with stimulated cycles.

It is also important to remember that gonadotrophins were—and still are—used to stimulate the ovaries without IVF. The aim of this 'hyperstimulation' is only to increase the number of mature eggs in the ovary. Without monitoring and left to natural fertilisation, hyperstimulation carries a great risk of multiple pregnancy, which, as we shall see throughout this book, may have complications for both the mother and her babies. Ovarian stimulation followed by IVF, however, does not just increase the chance of pregnancy but also gives the clinic a degree of control which natural fertilisation cannot. The clinic can select the best embryo after a few days of culture and transfer just one or two to achieve a singleton delivery.

Step by step in IVF

Today, 30 years after those first reports, most IVF treatment cycles still begin with a course of gonadotrophins to stimulate the ovaries, with multiple follicles generating multiple oocytes considered an optimal starting-point for the laboratory procedure. Individual variability in the ovary's response to stimulation is distinct, and as a result intense monitoring is required—and the extent of stimulation may be adjusted accordingly. Because ovarian stimulation aims to generate many eggs (and not the usual one as happens in a natural cycle), the patient's own endogenous hormone balance is disrupted, which may jeopardise the outcome. It's for this reason that complex stimulation regimens using multiple drugs are required.

Nevertheless, despite close monitoring and adjustments to the dose of gonadotrophins, some women may still respond with too few eggs, a phenomenon known as 'poor ovarian response', which in itself is associated with poor clinical outcome. In contrast, other women may exhibit many developing follicles (sometimes more than 20), and may end up with a dangerous health-threatening condition referred to as ovarian hyperstimulation syndrome (OHSS). The aim of ultrasound and hormonal monitoring, therefore, along with individualising each patient's drug dose, is to strike a balance between under and over response. Too few developing follicles could mean that the treatment cycle is cancelled, with no prospect of pregnancy; too many might raise the risk of OHSS—and cancellation for a more threatening reason.

Recent years have seen the development of stimulation approaches tailored to the characteristics of each individual patient, dependent on her age, type of infertility, and reproductive potential of her ovaries (as determined by hormone measurements). This has proved a major challenge, with only limited success achieved so far.

Once monitoring reveals that the follicles have developed into fully matured eggs, they can be harvested with the guidance of ultrasound through the vagina. This procedure is invariably unpleasant for the patient, not least because local anesthesia is only marginally effective.

The harvested eggs are then cultured in a dish in the laboratory and around 60–70 per cent will achieve fertilisation following exposure to sperm. These fertilised eggs, which first form their two 'polar' bodies and then at first division their two 'pronuclear' cells, are left to develop as embryos for between two and five days in the IVF laboratory. The embryos are stored in culture medium in an incubator to ensure their safe and well-supported development. Current opinion suggests that embryo quality may be better assessed after a longer period in culture—that is, until the embryos reach the blastocyst stage of their development at around the fifth day. However, to maintain optimum embryonic growth for five days requires optimum laboratory performance, and many questions remain about the impact of embryo culture on development potential.

So far, an embryo's quality has been almost exclusively assessed by its appearance under the microscope and rate of cleavage. However, because the ability of an embryo to implant in the uterus varies between 30 and 60 per cent, choosing the right embryo is critical to the outcome of IVF and ICSI. It's for this reason that historically multiple embryos have been transferred, to give the couple a better overall chance of pregnancy. However, the price to be paid for the transfer of multiple embryos is an increased risk of multiple pregnancy, which invariably occurs when more then one embryo implants. Today, in an almost worldwide initiative to reduce the rate of multiple pregnancies in IVF, few clinics transfer more than two embryos in routine cases; any supernumerary good quality embryos are stored in deep-freeze liquid nitrogen and left for later opportunities if the fresh attempt fails or for a second pregnancy.

Research in IVF today is focused on the quality and transfer of fewer embryos in this bid to reduce IVF's unacceptable rate of multiple pregnancy. Much progress has been reported in milder approaches characterised by the transfer of just one good quality embryo, ovarian stimulation with lower doses of gonadotrophins, and new molecular technologies to better assess embryo quality. The aim is to improve embryo selection, make possible the transfer of fewer embryos without compromising success rates, and reduce the rate of multiple pregnancies, complications and cost. Better embryo selection will also improve the chances of embryo survival with the new rapid freezing technologies available, and improve our understanding of the role of the uterus in embryo implantation.

The clinic takes control

The sudden and large-scale introduction of ICSI in the early 1990s would not only broaden the indications for IVF—now to provide an effective treatment for all forms of male infertility—but it also gave the clinic even greater control over the entire assisted reproduction procedure. Greater control, however, was not the aim of ICSI at the outset. Male infertility, caused by too few or even absent motile sperm cells in the ejaculate, was simply not treatable by conventional IVF. Some studies showed that as many as one-third of all cases of infertility were caused by a male factor, some of them beyond the reach of IVF. Many men were thus denied the chance of fatherhood—and many couples of parenthood—unless they turned to donor insemination or adoption. However, emerging techniques of 'micromanipulation', borrowed from veterinary medicine and dependent on microscopes able to magnify several hundred times, offered the first flicker of hope.

In 1988, at the National University Hospital of Singapore, Soon-Chy Ng and Ariff Bongso reported the world's first live birth following the transfer of an embryo fertilised by a single sperm cell delivered by hand, not by nature. Beneath the lens of a high-magnification microscope and held in place by a pipette 10 times finer than a human hair, a few sperms had been introduced beneath the outer layer of a human egg, in a procedure later known as sub-zonal insemination, or SUZI. One of the injected sperms had fertilised the egg, and in so doing had finally exposed a new possibility for the effective treatment of male infertility.

Within a year at least 10 clinics around the world were reporting their results with SUZI, including ours at the VUB in Brussels. And it was at the VUB that a young Italian gynaecologist, Gianpiero Palermo, found himself working on SUZI oocytes one afternoon in October 1990. In one of several procedures Palermo had inadvertently allowed the injection needle to penetrate beyond the membrane of the outer zona of the egg and into the inner cytoplasm. Palermo left this 'damaged' egg with the other SUZIs in the incubator overnight. And next day, along with some of the others, it had fertilised, in what was the first fertilisation from direct intracytoplasmic sperm injection.

An initial comparison of SUZI and ICSI fertilisation rates in VUB trials demonstrated the indisputable superiority of ICSI over SUZI—a fertilisation rate of 51 per cent in the former and just 17 per cent in the latter. As a result, the VUB abandoned SUZI in favour of ICSI in August 1992, confident that male factor infertility, so far one of reproductive medicine's greatest challenges, was now at last amenable to treatment. Until then, semen which contained fewer than a half million sperms was beyond the scope of IVF; now, just one was enough, and the whole miracle of reproduction lay—quite literally—in the hands of the embryologist.

Today, ICSI is the world's favoured method of fertilisation in assisted reproduction, even though reports from the world's various IVF registries show no difference in overall pregnancy or live birth rates between IVF and ICSI (around a 30 per cent per transfer ongoing pregnancy rate with each). However, there is still a steeply rising trend towards ICSI as a favoured fertilisation method. For example, European IVF monitoring figures for 2001 show that 49 per cent of all assisted reproduction cycles were with ICSI; by 2003 this proportion had increased to 55 per cent, and by 2005 to 63 per cent. In some countries—notably Germany (70 per cent), Italy (73 per cent), Belgium (75 per cent), Spain (83 per cent) and Turkey (97 per cent)—ICSI cycles far outweighed IVF, yet in others—UK, Netherlands, Denmark, Sweden—IVF remains the dominant technique.

This trend towards an ever increasing use of ICSI has been observed throughout the world. In Australia and New Zealand 59 per cent of all cycles used ICSI in 2005 and in the US almost 60 per cent. Commentators from data-collecting groups have

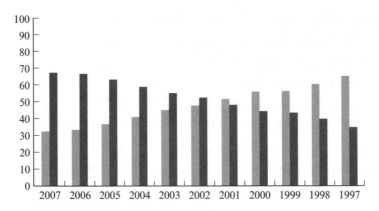

Figure 2 In just 20 years ICSI has become Europe's most favoured method of fertilisation in assisted reproduction. The most recent uptake data from ESHRE show that ICSI (pale grey bars) is now performed in more than 60 per cent of all assisted reproduction cycles, and twice as often as conventional IVF. In some countries the use of ICSI is almost 100 per cent.

been unable to explain such a dramatic increase in the use of ICSI—and certainly not by a proportional increase in male infertility, which was after all ICSI's original indication when the VUB pioneered its use in the early 1990s. There now seems to be more frequent use of ICSI in cases with unexplained and mixed cause infertility, but the trend seems mostly the result of professional preference—and a reflection of the fact that a specific cause of infertility is increasingly not investigated and not diagnosed.

Despite these contemporary trends, however, ICSI still has its most effective role to play in treating couples whose infertility is of male cause, and this would cover a range of indications from subnormal levels of sperm concentration or motility to complete azoospermia, where no sperm cells are evident in the ejaculate. And for men with these devastating conditions ICSI has finally made the prospects of fatherhood possible.

ICSI has also made the prospects of fatherhood excellent for men with cystic fibrosis, a genetic disease formerly associated with poor prognosis and childlessness. Indeed, even men without the symptoms of this lethal condition but who carry mutations of the causative gene have a one-in-four chance (according to Mendelian law) of passing on the disease to a child if their partner is also a carrier of the mutation. The responsible gene is known as cystic fibrosis transmembrane conductance regulator (CFTR), and carrier frequency in the Caucasian population—and in certain sub-populations such as Ashkenazi Jews—can be as high as one in 25. Around one in 2,500 newborn babies are affected, and almost all have a poor life expectancy of less than 40 years.

More than 30 years ago studies showed that nearly all men with cystic fibrosis were also infertile, while women with the same disease were not. It was thus found that in men cystic fibrosis not only caused a debilitating mucus abnormality in the lungs and pancreas but also a congenital absence of the passage which transports sperm from the testes to the urethra—known anatomically as the vas deferens. Before the introduction of ICSI, men with cystic fibrosis—notwithstanding the genetic risks to offspring—were simply unable to father their own children because no sperm cells were ever available in the ejaculate. Now, ICSI using surgically retrieved sperm cells—from the epididymis and testis—has revolutionised the prospect of fatherhood for these men, and provided a source of sperm cells which biology and genetics had formerly denied them. Success rates of around 50 per cent and higher are widely reported today.

It is a precondition of treatment, however, that both the male with cystic fibrosis and his female partner are assessed for mutations and polymorphisms of the CFTR gene. It is unavoidable that each child of a man with cystic fibrosis—as with all single gene defects—will be at least a carrier of the abnormal gene, but the actual risk of disease to the child will depend on the carrier status of the female partner. Only if the female partner also tests positive for a mutation of the CFTR gene (specifically, a gene known as delta F508) will the child be affected; having two copies of this delta F508 mutation, one inherited from each parent, is now known to be the leading cause of cystic fibrosis. Generally, having an identical pair of genes with the delta F508 mutation prevents the CFTR protein from fulfilling its normal function in cell membranes, thereby causing water retention in cells and the eventually lethal symptoms of the disease. It's for this reason that genetic testing and counselling are so important in these cases.

Before the late 1990s a termination of pregnancy following positive prenatal testing for cystic fibrosis was the only way to prevent the birth of an affected child. But since then, the technique of preimplantation genetic diagnosis (PGD) can provide the same accuracy of results within a few days of fertilisation, and not after two or three months; PGD has thus provided a reliable alternative to any future prenatal testing and 'therapeutic' termination of pregnancy.

PGD allows a single cell biopsied from embryos created by ICSI to be 'diagnosed' for the presence of genetic disease. This diagnosis is usually derived from results of a mutation-specific test on the biopsied cell. If the test shows that the mutation is not present, the embryo from which the biopsy was taken can be transferred as healthy and unaffected. Conversely, cells which show the mutation will represent an affected embryo, and will not be transferred.

Thus, not only has ICSI made biological fatherhood possible for men with cystic fibrosis, but PGD, as we propose later, has also made it safe. Nowhere, therefore, has assisted reproduction and its attendant science been better able to 'design' a healthy baby than in the case of cystic fibrosis: ICSI made fertility possible, PGD has removed the risk of disease. Both were achievements of enormous magnitude, though safety, through the application of PGD, can only now be claimed when the full range of cystic fibrosis gene mutations are tested for.

The healthy child born to a man with cystic fibrosis represents a triumph of reproductive science and medicine over the vagaries of nature. With other similarly visible successes over the challenges of nature, it's no surprise that the number of couples turning to assisted reproduction as a solution to their infertility continues to grow. In Denmark up until 2010, where IVF and ICSI had been publicly funded and easily accessible in hospital clinics, as many as 6 per cent of all children born were conceived by assisted reproduction. In every nursery classroom in Denmark there are now two, three or more IVF and ICSI children. So today, with a variety of treatments added to the basis of IVF—such as egg donation, egg, sperm and ovarian tissue freezing to preserve fertility, donor insemination in lesbian couples and single women, surrogacy—there is no reason why anyone should do as our grandparents did and simply accept infertility and the shortcomings of reproduction as nature's way. Almost all causes of infertility are now amenable to treatment, and the technologies which have made that possible are not the blueprint for some Orwellian brave new world, but the means to overcome the flaws of nature and deliver a healthy baby. The science of designing babies is not to enhance the genes of popular appeal, but to overcome the sometimes tragic consequences of postponed childbearing or defective genes and to repair the breakdown of reproductive biology.

What Couples Want and How We Deal With It

In the dark ages of 1980s California, a millionaire who had made his fortune protecting human eyes (with shatter-proof spectacle lenses) turned his attention to protecting the human race. Like some villain in a Superman comic, Robert Graham planned the creation of a super race which would, he hoped, reverse the decline of America's gene pool. 'Early in my life it dawned on me that bright, desirable citizens weren't reproducing themselves,' he explained in a 1983 interview. Graham's saviour plan for the human race was a sperm bank whose only donors were proven geniuses—Nobel laureates, rocket scientists—and great artistic achievers. Intelligent selection, he proposed, held the key to 'a new level of being'.

It was no surprise that the crackpot scheme floundered. Reportedly, only three Nobel laureates signed up, and one—the only name ever identified—was, at 70 years old, probably past his best. Moreover, when the scheme—known as the Repository for Germinal Choice—was exposed by the *Los Angeles Times,* there were justified cries of protest that this was little more than eugenics masquerading as medicine. Indeed, Graham's one known Nobel sperm donor, the physicist William Shockley, had long been identified as a racist and white supremacist.

It was activities like this that first raised the spectre of 'designer babies'. The idea of 'designer jeans'—by which a fabric as mundane as denim received the personalised attention of branded haute couture—had first surfaced in the late 1970s and had quickly been applied to handbags, shoes and sunglasses. So why not babies? Thus, the first designer babies were not so much designed at conception, but more kitted out in clothing and buggies designed solely to impress and put a stamp of character on an otherwise everyday toddler. But it took no great leap of imagination to extend the designer concept to the genius babies set to emerge from the Repository of Germinal Choice.

Today, the science of genetic heritability remains inexact, but the concept of eugenics, as exemplified in the ethnic cleansing of Nazi Germany, is unequivocally abhorrent throughout the world and defined as such in the United Nations' Universal Declaration on Human Rights. So when a California fertility clinic claimed on its website in 2009 that personal traits—such as hair and eye colour—could be tailor made in their babies, the worldwide uproar of protest was still as great as ever. Critics claimed that this kind of conceptual discrimination was not so much eugenics but more a kind of medicine which deselects embryos with unwanted genes or gene variants not for reasons of health but of cosmetics. Single nucleotide polymorphisms (SNPs) associated with hair and eye colour in Icelandic and Dutch populations had been reported in 2007 from genome wide association studies, so the means to identify such gene variants in embryonic cells (at least in theory) did and does exist; however, what's not yet current is the universal will to apply such a technology for non-medical reasons.

Nevertheless, the will to apply some degree of cosmetic matching is evident in the very common fertility procedures of egg donation and donor insemination—even if choosing a sperm donor is far less ethically and technologically demanding than selecting an embryo. Before the widespread availability of ICSI in the 1990s, donor insemination was the only effective treatment for couples whose infertility was caused by a sperm problem in the male partner. And then, as today, there was a deliberate effort to match the basic characteristics of the sperm donor with those of the recipient patient.

These characteristics are first and foremost related to ethnicity, and then to culture, religion, physical traits (such as height or complexion) and background. Clinics and sperm banks in the US, for example, where federal regulations strictly control the testing and screening of sperm donors for infectious diseases, are otherwise happy to reveal quite personal information about their donors in the belief that such knowledge will make the recipient couple a little more at ease with their 'anonymous' donor. Fairfax Cryobank in Virginia, for example, lists donor xxx as '100% Vietnamese...hard working and determined...shy at first but very polite'. Indeed, there are numerous examples—and one assumes a great many more than have been publicised—of donor sperm recipients doing as much as allowed to ensure that the donor meets their personal requirements. There were unconfirmed reports (in the ever inventive British press) in 1998 that Hollywood actress Jodi Foster had been inseminated by 'a tall, dark, handsome scientist with an IQ of 160'.

The genetic contribution of sperm to the final composition of a baby is, as that of the oocyte, 50 per cent. The father provides one copy of each of 23 pairs of chromosomes, and the mother the other copy. But which genes on the chromosome make that contribution is largely indeterminate. Nowhere is this better illustrated than in the breeding of the racehorse, a self-contained species whose complete ancestry can be tracked back in scrupulously kept records to just three stallions imported to Britain in the early eighteenth century. Yet today, after generations of research and pedigree analysis, breeders can offer few better clues to breeding a champion than 'mate the best to the best—and hope for the best'.

Racehorse breeding, claimed the celebrated Italian thorough-bred breeder Federico Tesio, is the perfect demonstration of Mendel's laws of genetics. Mendel had actually experimented with peas, not racehorses, and had crossed a tall variety of pea with a dwarf variety, to produce hybrids. The hybrids turned out to be tall, but when these hybrids themselves were crossed with each other the result was a mix of tall and dwarf varieties—in a ratio of three tall to one dwarf. Mendel thus concluded that nature has a tendency to return a breed to its original state. And it was this tendency, Tesio reasoned, which explained why one racehorse might be brown in colour and its full brother (same father, same mother) might be chestnut. It would also explain why one might run fast, and the other could hardly pass a donkey.

Like the racehorse breeder, Robert Graham in the genius sperm bank of his Repository for Germinal Choice sought to 'breed the best to the best', but the outcome has not been catalogued. More than 200 children are said to have been born as result of the sperm donor experiment, but few have come forward. From those who did, there are anecdotal reports of artistic, athletic and scholastic achievement, but, as one journalist reported of one, only 'a smart but not supernatural college student' was the result.

Certainly, from the studies of how much intelligence is 'nature' and how much 'nurture', it seems unlikely that such complex traits as IQ can be inherited in any complete predetermined way. The contribution of the genotype (the genetic make-up of an individual) to the phenotype (the visible characteristics of an individual as determined by genes and environment), therefore, seems limited, with most studies calculating an 'heritability' value of between 0.4 and 0.8 (on a scale of 0 to 1); adult height, however, has a heritability estimated at 0.80, when a similar environmental background is present.

But this is an age of consumer choice, and few sperm banks today (or even IVF clinics offering egg donation) in consumer driven societies like the US would promote their sperm donation services without a clearly itemised choice of donor.

In our experience in Europe and almost without exception, the patients we see today coming to our clinics for treatment have only one thing in mind, and that's no more than the birth of a healthy baby of their own. We have encountered few who discriminate positively, and their hopes are pinned exclusively on the dream of a baby, their life fulfilment in the form of a happy and healthy family. They rarely think that anything can go wrong, rarely contemplate the psychological demands of treatment, and will often spend more than they can afford to realise this dream. For them, infertility treatment represents an enormous life choice which is dependent on a complex catalogue of expectations and a determination to commit their deepest emotions (and often great expense) to the fulfilment of this wish. Thus, what they—and we as doctors—are designing is not so much a tailor made baby with its own 'designer' traits, but more the 'happy family' which life so far has denied them.

Their motivation to achieve this is substantial, and no better illustrated than in the lengths infertile couples will go to (in terms of effort and stress) to be treated as they feel necessary. We see it particularly in the recent trend of 'cross-border' reproductive care where patients denied a treatment in one country will travel to another where that treatment is available. In France, for example, donor insemination or IVF treatment is not allowed by law in lesbian couples, so there is a steady trail of women in same-sex relationships travelling from France to Belgium, where such treatments are permitted.

Similarly, egg donation—the only viable treatment for women with a premature menopause—may be difficult in countries with a short supply of donor eggs and long waiting lists for treatment. In the Netherlands, for instance, women requiring egg donation can only be treated if they find their own suitable donor. No financial incentives are allowed, and no-one is willing to donate to an unknown recipient. In Spain, however, where egg donors are better compensated than in some countries, the supply of oocytes is more plentiful and treatment more immediately available. And newly introduced cryopreservation techniques have now made oocyte banking just as efficient as sperm banking, with an even greater abundance of supply in some countries. So why wait two or three years in Britain, for example, when the same treatment is ready and waiting—and with equally high success rates— in Barcelona? Websites for clinics in Spain, Ukraine, Russia and Cyprus readily announce 'no waiting lists' and 'donors always available'. There are even 'fertility tourist' companies able to arrange flights, hotels and clinics.

A pilot study performed by the European Society of Human Reproduction and Embryology (ESHRE) in 2009—the first of its kind—found a considerable flow of patients crossing borders between most European countries for infertility treatment. During a one-month period, data were collected from clinics in six European countries (Belgium, Czech Republic, Denmark, Slovenia, Spain and Switzerland), and from patient questionnaires about age, country of residence, reasons for travelling, treatments received, clinic selection, and reimbursement. An extrapolation of the data by ESHRE gave an estimate of at least 20,000 cross-border treatment cycles now happening in these six countries each year.[1] Similar patterns—though on a smaller scale—have been reported from North America, by US citizens looking for less expensive treatment in Canada.

In the European study almost two-thirds of the patients surveyed came from four countries, the largest numbers from Italy (32 per cent), Germany (14 per cent), the Netherlands (12 per cent) and France (9 per cent). In total, people from 49 countries crossed borders for fertility treatment. The main reason for travelling was to avoid legal restrictions at home, particularly for German, Norwegian, Italian and French patients. Difficulties of access to treatment were cited more by patients from the UK than those from other countries. It was no mere coincidence that both Italy and Germany had at the time some of Europe's most restrictive laws on embryo selection and embryo freezing.

For most of these patients travelling abroad for treatment, their cross-border care invariably represents a huge investment in time, money and emotion, yet the cost and disruption caused in search of treatment seem modest when seen alongside the psychological stress which IVF today necessarily implies. Indeed, one infertility counsellor has compared the high levels of stress seen in some fertility patients with those in cancer patients;[2] similarly, a study reported at the annual meeting of ESHRE in 2008, which assessed psychological factors in infertility by the use of language, found the same intensity of emotions only in oncology patients.

More specifically, a study from 2004—the first to use structured psychiatric interviews instead of self-reported measures—found that 40 per cent of infertile women immediately before their first clinic visit were depressed, anxious or both.[3] It was no surprise, therefore, that a more recent study from our group in the Netherlands identified 'emotional distress'—ahead of 'poor prognosis'—as the main reason for dropping out of an IVF treatment programme. The same study—a prospective analysis of almost 1,400 IVF patients in Utrecht—showed that half the total drop-outs (45 per cent) had actually given up even before their treatment had started. The investigators' first explanation for such an early discontinuation of treatment was the patients' own decision to 'reject fertility treatment in general', and second, 'relational problems'. However, throughout the entire study period of three IVF cycles the greatest reason for discontinuation was 'emotional distress'.[4] Given the intensity of emotions which infertility itself invariably causes, it is not surprising that its treatment precipitates comparable reactions, ranging from the intense joy of parenthood to the depths of despair when treatment fails.

A singleton pregnancy, or twins?

With stress levels so high (as well as costs), it's no surprise that many, if not most, couples coming to a first consultation for IVF would be quite happy to have twins. A ready-made family in one go, two bundles of fun at only half the price and with only half the worry. The facts bear this out. A study of all IVF clinics in the Netherlands found that 36 per cent of couples positively wanted twins, and almost the same proportion thought it was their right to decide how many embryos should be transferred.[5] They were also aware that transferring just one embryo—and therefore avoiding the risk of twins— might lower their chances of success; so why raise the odds?

The evidence also suggests that many IVF doctors are happy to go along with their patients' wishes, despite the unequivocal wisdom that twin pregnancies carry greater health risks to the mother and her babies than singletons. Thus, even though IVF monitoring data (for 2007) have shown that, for the first time since records began, the overall twin delivery rate after IVF and ICSI in Europe has now fallen below 20 per cent, with no quadruplets reported, the majority of cycles are still reliant on the transfer of two embryos. Ten years earlier, in 1997, Europe's multiple delivery rate was 30 per cent, with triplet and quads at 4 per cent. So there is a definite trend in favour of singletons and single embryo transfer—but nevertheless, two embryo transfers are still the norm. Indeed, in some countries

ESHRE data for percentages of singleton, twin, triplet and quadruplet deliveries from IVF and ICSI procedures in Europe 1997–2007. There is great variation in rates per country, with, for example, a multiple delivery rate of 5.3 per cent recorded for Sweden and 23.5 per cent for the UK.

	1997	2003	2007
Singleton	70.4	76.7	78.2
Twin	25.8	22.0	20.5
Triplet	3.6	1.1	0.8
Quad	0.15	0.08	–

of Eastern Europe more than 50 per cent of all transfers in 2007 were still of three or more embryos. And even in Germany, France and the UK, double embryo transfers accounted for more than 60 per cent of all procedures. Only in Belgium, the Netherlands and the Nordic countries did single embryo transfers represent more than 50 per cent of all procedures.

Even more alarmingly, the latest figures from the US (for 2008) show that only 5 per cent of all IVF and ICSI cycles in younger patients were single embryo transfer; and in this same age group (under 35s) one in three of all deliveries were twin or higher order multiples, which means that more than 50 per cent of all children born from IVF in the US are from multiple pregnancies.

Some opponents of single embryo transfer in the US, where the fertility 'market' is more commercially driven than in Europe, have insisted that, 'as a universal medical principle', patients are entitled 'to maximal professional efforts towards their desired outcomes in the safest, quickest and most cost-effective ways'. And in their estimation this has invariably meant putting the 'benefits' of a twin pregnancy before its risks. One such American iconoclast, Norbert Gleicher from the Center for Human Reproduction in New York, even developed a statistical model which showed that for those patients wanting more than one child, 'twin deliveries represent a favorable and cost-effective treatment outcome that should be encouraged, in contrast to the current medical consensus'.

But for most IVF clinics—and the agencies which regulate and guide them—twin deliveries are one outstanding area of assisted reproduction where the wishes of the patient are not necessarily those of the doctor. For more than two decades the professional organisations involved in assisted reproduction have expressed concerns, initially raised by obstetricians, about the escalating rate of twin pregnancies in IVF; a review of all IVF births in Denmark, for example, confirmed that the twin birth rate doubled between 1970 and 1996, much of it accounted for by IVF. Behind these fears lies the indisputable fact that the most striking health risks found in neonatal infants are associated with multiplicity. And even though the prevalence of twin pregnancies in IVF is now slowly beginning to shrink, in both Europe and North America, the fact remains that one in five of all deliveries in Europe (and one in three in the US) are still twin and that almost one in three of all IVF and ICSI babies are born as twins. We reviewed the risks of multiple pregnancies for *The Lancet* in 2005, noting indeed that the rise in prevalence of twin, triplet and higher order multiple

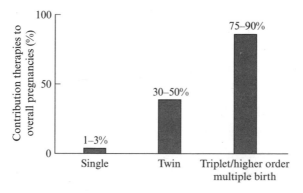

Figure 3 The contribution of assisted reproduction to the total numbers of single, twin and higher order births

births was mainly the result of ovarian stimulation for assisted reproduction.

A few years ago researchers in Denmark, where every individual has a unique identification number which can be cross-checked against different databases, analysed the country's birth registry to find every woman who had given birth to twins between 1995 and 2000. This cohort was then checked against Denmark's IVF registry to identify every single twin birth resulting from assisted reproduction and provide a comparative group with those spontaneously conceived—a total of 3,438 IVF/ICSI twins and 10,362 control twins.[6] When these two groups were compared in Denmark's patient registry, several striking differences emerged. First, both the average birth weight and gestational age were significantly lower in the IVF/ICSI twins than in the control twins, and the risk of delivery before 37 weeks significantly higher. In addition, the rate of Caesarian sections was higher in the IVF/ICSI twins (53 against 43 per cent), as was admission to a neonatal intensive care unit.

Similarly, a more recent study of all births in Sweden between 1982 and 2007, which compared 1,545 pairs of IVF twins with 8,675 spontaneously conceived twins, also found the risk of preterm delivery before 32 weeks significantly increased among the twins born after IVF.

Such trends in prematurity matter because many studies have shown beyond doubt that infants born before term are at risk of developing a wide range of health complications, not just those which can be managed in the neonatal unit but long-term ones too. For example, the incidence of cerebral palsy is higher in infants of very low birth weight

Complications related to multiple pregnancies

In the mother	In the child
* Miscarriage	* Prematurity
* Anaemia	* Low birth weight
* Pre-eclampsia	* Morbidity (even mortality)
* Gestational diabetes	* Malformation
* Growth retardation	* Cerebral palsy
* Caesarean section	* Disability
	* Learning difficulties
	* Adult health risks (Barker hypothesis)

(under 1,500 grams) than in those of normal birth weight. Recent research indicates that rates of cerebral palsy are falling, even in very low birth weight babies, but the association still remains. Indeed, a very recent study from Denmark of more than 500,000 children found that those conceived by assisted reproduction ran almost twice the risk of cerebral palsy as those conceived naturally.[7] The investigators in this study noted that 'this increased risk of cerebral palsy in children born after assisted conception, and in particular IVF, is strongly associated with the high proportion of multiplicity and preterm delivery in these pregnancies'.

Other complications associated with multiplicity have been found to affect both the mother and her babies. Maternal rates of pre-eclampsia, gestational diabetes and growth retardation have all been found higher in women with multiple pregnancies, as are rates of Caesarean section and miscarriage. Babies born from multiple pregnancies have also been associated with slower developmental progress (which has even continued into high school). Compared with singleton births, perinatal mortality rates are at least four-fold higher for twins—and at least six-fold higher for triplets. And in twin and higher order multiple births studies have found the risks of prematurity are enhanced between 7 and 40-fold, and of low birth weight 10 and 75-fold. It is against the background of such findings that preterm delivery has been described as the most important risk factor for neonatal mortality and morbidity—and in turn why multiple pregnancy defined as the major health hazard of IVF and ICSI. Low birth weight has even been associated with numerous adverse conditions much later in life, with some studies suggesting a link with cardiovascular disease, diabetes and impaired longevity.

Thus, the impact of the deleterious effect of multiplicity on assisted reproduction is such that most safety studies of IVF find that its only major complication is multiple pregnancy, not the IVF or ICSI procedure itself. This was illustrated in a long-term study of ours at the VUB in Brussels, where we have been following a cohort of ICSI babies since birth, with the oldest in our most recent study eight years of age. At the time, we had overwhelming reasons to begin such a study, for when we were developing our ICSI programme in the early 1990s it was a matter of some concern to us that we could achieve the same pregnancy rates with very poor sperm as with normal samples. Would this apparent paradox be reflected in the outcome of the pregnancies? We were also very sensitive to the fact that—for the first time in reproductive medicine—human sperm for fertilisation was now being selected in the laboratory by an embryologist, and not as nature intended. And we were worried too that the sperm cell had to be invasively injected into the female's oocyte, and not left to find its own way through the outer zona and into the cytoplasm. Thus, with so much potential for harm and to establish the database which would prove (or disprove) the safety of ICSI, we set about the huge task of developing a pregnancy and children follow-up programme.

Now, after many analyses and more than two decades of investigation, we can confidently conclude that most of our worries were misplaced, that the rates of mild and major malformations in the ICSI pregnancies and children are no different from those of the children conceived by conventional IVF—or from those found in the general population. In fact, the only difference we did find was, when the data were analysed in the presence of triplets or even twins, long-term development outcome seemed worse because of prematurity. Any problems, therefore, were not related to the technique of ICSI (despite sperm selection by the embryologist and the injection technique) but to the occurrence of multiple pregnancy.

We should add here that the risk of twins and triplets is not confined to IVF and ICSI. The use of fertility drugs to induce a spontaneous ovulation (a technique known as ovulation induction) or multiple ovulations with or without insemination have also been justifiably implicated in a raised risk of multiple pregnancy. Although both techniques use fertility drugs, they differ in both the type of patient treated and in the objective of treatment. The former aims to induce a single ovulation in women who (usually for hormonal reasons) are unable to ovulate. The latter, often referred to as 'hyperstimulation', is usually used to treat women with normal menstrual cycles and aims to induce the development of many follicles, as in preparation for some insemination procedures. The risk of multiple pregnancy is especially high when the effect of treatment on the development of eggs is not adequately monitored.

Indeed, some of the most shameful cases of high-order multiplicity have been in patients receiving fertility drugs for non-IVF ovarian stimulation, which, if properly monitored, need rarely have resulted in high-risk multiple births. Ovarian stimulation in the context of IVF, however, need not be linked to multiplicity, because the number of embryos transferred can be limited. Conversely, the number of dominant follicles induced by hyperstimulation in non-IVF treatments is beyond our control, and careful monitoring by ultrasound is necessary.

However, the octuplets born in January 2009 to a 33-year-old single mother in California were conceived by IVF—despite the guidelines of the American Society for Reproductive Medicine (ASRM) recommending that no more than two embryos be transferred in women of this age. All the babies were born prematurely and delivered by Caesarean section—and all required intensive care and intravenous feeding. According to reports, the California Medical Board accused the mother's doctor of 'gross negligence', complaining that his treatment was 'beyond the reasonable judgment of any treating physician'. Press accounts also suggest that the single mother, Nadya Suleman, who already had six children, was later counselled on options to reduce the pregnancy by a process known as embryo (or selective) reduction, but refused.

Embryo reduction, an intervention by which 'surplus' embryos are aborted, usually occurs at between seven and 12 weeks of the pregnancy. The embryos are visualised by ultrasound and either removed by aspiration, (most commonly) through the vagina, or terminated by chemicals to reabsorb into the woman's body. The procedure is not without risk to the other remaining embryo(s), and has thus been mainly applied in cases when three or more embryos implant in the uterus, and when the hazards from the multiple pregnancy are undoubtedly greater than those from the embryo reduction.

For many patients—and in many countries—embryo reduction does pose ethical problems. It was Ms Suleman's right to refuse the offer of embryo reduction, but we agree with the California Medical Board that her doctor had no right to transfer 'eight' (or six in some reports) embryos, even if that was his patient's wish. The transfer of so many embryos went disastrously beyond the guidelines of the ASRM, and, as the California Medical Board stated in its charges, Ms Suleman's doctor 'failed to exercise appropriate judgment and question whether there would be harm to her living children and any future offspring should she continue to conceive'.

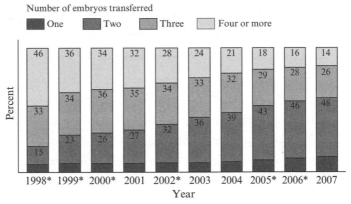

Figure 4 Despite a declining trend, 14 per cent of all embryo transfers in the US in 2007 were of four or more embryos. Single embryo transfer still accounted for only 12 per cent of all procedures.

These same headline cases, therefore, also underline the unwelcome truth that the risks of prematurity, low birth weight, Caesarean delivery and neonatal intensive care all increase with the higher-order of pregnancy. Statistically, triplets have a far worse prognosis than twins. Yet the fact is that there are still many countries of the world where three or more embryos are routinely transferred to IVF patients. Data from the Centers for Disease Control (CDC) in the US show that as recently as 2007 four or more embryos were transferred in 14 per cent of all first cycles of IVF or ICSI. The majority were in women over the age of 35, but this pattern did not prevent the fact that 4 per cent of all IVF and ICSI births in the US in 2007 were 'triplet-or-more'. By our calculation, therefore, 1,443 of the 36,079 pregnancies

41

recorded in the US were at least triplet in 2007, and exposed to unnecessary neonatal risk. In reporting the results, the CDC added: 'It is important to note that twins, albeit to a lesser extent than triplets or more, are still at substantially greater risk for illness and death than singletons...Both percentages of twin and triplet-or-more births remain significantly higher for [assisted reproduction] births than for births resulting from natural conception.'

Nor are high-order births confined to the US. ESHRE, in its monitoring of European IVF and ICSI, provides evidence of declining triplet rates, but in 2007, 11 per cent of all deliveries in Latvia were still 'triplet +', and the figure in Serbia was 3 per cent. Even in Greece, almost 70 per cent of all procedures transferred three or more embryos.

It is for these reasons that the world's two leading professional organisations in reproductive medicine, the ASRM and ESHRE, have each named multiple pregnancy as the most serious complication of IVF and ICSI, the former somewhat more forcibly than the latter. We agree with this view, and it is our opinion—as well as a theme of this book—that safe IVF, as demonstrated in the birth of a healthy singleton baby, is today the overriding motivation for 'designing' a baby by assisted reproduction. The debate on multiple pregnancies, and the fact that many infertile couples are still emotionally inclined towards twins, suggests that the wish of the doctor and the wish of the patient are not always one and the same. However, studies have also shown that counselling before treatment and an IVF programme which comprises a fresh cycle of treatment followed by the cryopreservation and subsequent transfer of spare embryos, go a long way towards an agreed course of treatment based on informed choice and the availability of good scientific evidence.

Single embryo transfer

Both ESHRE and ASRM position statements name elective single embryo transfer (eSET) in appropriately suitable patients as the only effective way to diminish the high rate of twin pregnancies. And this too is not always a matter of agreement between doctors and patients, raising the difficult ethical question of who has the greatest right to decide how many embryos to transfer: the doctor, the patient or even the regulatory authority? Indeed, in Sweden and Belgium, IVF clinics are compelled by regulation to transfer just one embryo in a first IVF or ICSI cycle, and in the UK the regulatory body, the Human Fertilisation and Embryology Authority (HFEA), has also set a target for clinics to reduce their multiple pregnancy rate to 10 per cent by pursuing an eSET policy.

Resolution of this contentious debate is compounded by two issues: first, whether treatment is funded by the patient or by the state; and second, the fact that success rates have been found in most studies to be lower when one embryo is transferred than two. So the argument goes that when IVF is offered in an exclusively private market—as in the US—the patient paying the bill has a right to expect the highest chance of a pregnancy and live birth in a single cycle of treatment, and that this can only be achieved with the transfer of two or three embryos.

The latter argument is also compounded by how 'success' in IVF is measured: whether the outcome of treatment is evaluated in a first treatment cycle when 'fresh' embryos are transferred, or in a programme of treatment which includes the initial fresh cycle and the subsequent transfer of frozen embryos. When IVF success is measured in the context of all embryos created from a single stimulation, cumulative birth rates are comparable in both single and double embryo transfer programmes. In the end, all available embryos will be transferred; however, in the former each embryo is transferred just one at a time.

There is now powerful evidence from studies in Finland and Belgium that cumulative delivery rates from eSET are just as high as those achieved from fresh cycles transferring two embryos. A large retrospective study from Finland reported a *cumulative* live birth rate per started cycle of 42 per cent in its recent eSET treatments, but only 37 per cent in the pre-eSET treatments.[8] The cost of delivering a healthy baby in the eSET programme was also found to be less expensive than in the pre-eSET period. By shifting their outcome measures to cumulative results per treatment cycle started, the Finnish investigators were able to recommend that 'eSET with cryopreservation is more effective and less expensive than double embryo transfer, and should be adopted as treatment of choice in IVF/ICSI'.

Similarly, in explaining why the UK's adoption of eSET was lagging behind most countries of Europe and Australia, the British Fertility Society proposed in 2008 a greater use of eSET to help clinics reduce their multiple pregnancy rates from an average of 24 to 10% over the next three years, as demanded by the HFEA, without compromising live birth rates. The BFS recommended that eSET plus subsequent frozen embryo transfer can be as effective as double embryo transfer in patients who are carefully selected - notably, women under 37 years of age and having a first cycle of IVF or ICSI (87 per cent of the UK's multiple pregnancies occur in a first cycle of treatment).

Guidelines and regulations

Few physicians working in fertility would disagree with our opinion—especially in the light of the California case—that the patient's hopes from treatment are not always those of the doctor, or that the doctor's practice always conforms with the guidelines. Thus, while almost all fertility doctors would refuse to transfer six embryos (as was reportedly done in the case of Ms Suleman—two of the embryos transferred were claimed to have produced identical twins), there may be more sympathy for a request to transfer three, or even four, especially in an older patient. Indeed, as we shall see in chapter 7, some physicians will interpret guidelines as specific circumstances allow, thereby applying a freedom of judgement which legislation would make impossible.

Some people, therefore, because of guidelines and legislation, are simply denied the treatment—whatever it may be—which best gives them the chance of a family of their own. Nor do these regulations just apply to embryo transfers and multiple pregnancy. Even within the confines of Europe there is a diversity of regulation governing assisted reproduction which removes any notion of harmonisation and is most evidently reflected in the recent patterns of fertility tourism.

These regulations impose restrictions on patients and on treatments. In Germany, for example, embryo freezing is banned by law, so the concepts of eSET, cumulative treatments, and cumulative birth rates are more difficult to realise. There are also different requirements throughout Europe on the identification of sperm and egg donors, which certainly has an impact on their number. In Britain, for example, the law on gamete donation was changed in 2008 such that donors would no longer be able to remain anonymous. The change of law now means that anyone conceived by donor sperm, eggs or embryos has, at the age of 18, the right to ask the HFEA for information about the donor. The new law also means that fertility clinics must now demand identifying—as well as non-identifying—information from all their donors.

This change in regulation may well be responsible for a decline in the number of UK sperm donors, which the British Fertility Society has recently described as 'critical'; figures from the HFEA, show that the number of women having donor insemination treatment in 2007 was 16 per cent lower than in the previous year; no more than 384 men throughout the country registered as sperm donors in 2008. In Spain and many other European countries, gamete donors can remain anonymous, and this is one reason why supplies of sperm and eggs appear more plentiful in these countries.

There were also changes to the UK's embryology laws in 2009 when single and lesbian women were granted the same rights to fertility treatment (and parenthood) as heterosexual couples. The UK shifted its overriding legal principle from 'the need for a father' to the concept of 'supportive parenting', and the new Act now recognises single women and lesbian couples as legitimate parents of children conceived with donor gametes. Sperm donation is the usual treatment in such cases, but there are many single women and those in same-sex relationships now turning to egg-sharing IVF, a treatment recognised as legitimate by the UK authorities for more than a decade. Egg sharers donate a proportion of their IVF eggs to another patient in exchange for reduced payment, with good results for both sharers and recipients. However, egg-sharing is not allowed, or is at least discouraged, in many countries because of the suggestion that subsidised treatment may be a form of inducement. There have been similar reservations expressed about the 'money back warranty' arrangements offered by some clinics. IVF patients have their costs returned if treatment fails, whereas those who become pregnant have to pay.

However, the most severe legal restrictions in IVF were seen in Italy in 2004 when the now infamous Law 40 was approved by the Italian parliament. Before then, Italy, unlike all other countries of the old Europe, had no legal provision for assisted reproduction. Practice had been left to self-regulation, respect for life and common sense, but commentators, mesmerised by the high-profile cases of postmenopausal pregnancies or claims that a human clone was imminent, had also described Italy as 'the wild west' of IVF; the new legislation would bring Italy into line with the rest of Europe.

Law 40, which had the support of the Catholic church and had overcome a national referendum on its way to the statute

book, outlawed embryo freezing and required clinics to fertilise no more than three oocytes, all of which had to be transferred. The result was a decline in live birth rates (in certain
patient groups) and an increase in multiple pregnancies (in
others). Other treatments involving embryo selection, or even
preimplantation genetic diagnosis to remove the possibility of
inherited disease in high-risk couples, were also banned. Such
measures no doubt explain why cross-border movements were
most prominently recorded from Italy in the ESHRE study,
reflecting the likelihood that legal restrictions in one country
are the main reason for reproductive tourism to another.

In 2009, in the wake of several legal challenges in lower
courts, the Constitutional Court of Italy finally declared parts
of Law 40 unconstitutional and removed the legal requirement
for clinics to fertilise a maximum of three oocytes in a single
treatment cycle. The Court also agreed that patient health is
paramount and thus declared embryo freezing constitutional
and justified if determined by the health of the patient. This latter ruling, observers suggest, has precedents in Italy's abortion
law where, under 1978 legislation, termination of pregnancy
is allowed if performed for the sake of the woman's health,
whether psychological or physical.

A similar requirement to protect the embryo lies behind
Germany's IVF laws, which continue to make embryo freezing
illegal, unless it can be shown that freezing is an emergency
measure undertaken for the sake of the mother's health. Thus,
the routine freeze-storage of fertilised eggs in IVF can only be
performed at the very earliest stage of their development—at a
stage of first division when just two pronuclei are evident.

Germany is also one of several European countries (Norway and Switzerland are other examples) to ban egg donation. Egg donation from an anonymous donor is banned in Finland, the Netherlands, Sweden and the UK, but is allowed if the donor provides identifying information. Conversely, egg donation from a woman whose identity is known is banned in Czech Republic, Denmark, France, Portugal, Slovenia and Spain. There are also ethical and legal differences in the acceptability of 'intrafamilial' gamete donation involving family members as the gamete donor. Such donation is illegal in some countries, although a recent position paper from ESHRE has concluded that gamete donation from family members may be acceptable provided that counselling has assessed any 'psychosocial and medical risks'.

Despite this diversity of regulation in Europe, there is nevertheless an underlying consistency of intervention which makes the treatment of infertility more heavily regulated than any other field of medicine. In many cases, these interventions seem to conspire against the right of a couple to have a child— or at least to raise a conflict between their legal rights and their human rights. It was this very conflict which lay behind the legal challenges to Italy's Law 40. But despite such regulatory diversity—or conflicts with human rights—there is also some consistency in the public view that the many techniques of assisted reproduction do require some kind of regulation, whether statutory or self-determined by guidelines.

In Europe especially the trend in regulation is more towards statutory control than to self-regulation by guidelines; Belgium, Czech Republic, Finland, France, Greece,

Hungary, Italy, Norway, Portugal, Spain, Sweden and the UK have all introduced new (or modified) legislation in the past few years, suggesting an ever stronger legislative hand, and most countries in Europe now have some form of statutory governance for assisted reproduction. The legislative trend in Europe, however, is in marked contrast with other populous countries of the world; India, Japan and the US still have no statutory requirements in place and remain reliant on self-determined guidelines. The hot spots for legislation—as recent UK changes reflect—are the marital status of the patient, the number of embryos transferred in each treatment cycle, embryo freezing, egg and sperm donation (with or without anonymity), and the genetic analysis of embryos by preimplantation genetic diagnosis (PGD). Even more controversial are sex selection for non-medical reasons, treatment of lesbian women, or pregnancies in postmenopausal women, all of them challenges for the medical ethicist which pitch the autonomy of the patient against the responsibilities of the doctor and state.

Moreover, a patient's rights to treatment can be as much tested by local guidelines as by legislation. There is continuing debate as to whether it is reasonable to deny treatment to women over a certain age (40, for example) or over a certain body weight. Another position paper from ESHRE has tackled this thorny question of withholding treatment for lifestyle reasons and concluded that obese and smoking women should make 'a serious effort' to cut back before IVF treatment is considered. But in the case of women 'used to more than moderate drinking' who are unwilling or unable to minimise their consumption, doctors 'should refuse treatment'. The paper took its position from an ethical perspective, but based its recommendations on the clinical effects of lifestyle paradigms on fertility (in natural and assisted reproduction), pregnancy complications and the welfare of the child. The central issue, as the authors readily noted, is where to draw the line in respect of the doctor's and patient's rights and responsibilities.

First, a background review of the 'facts' leaves little doubt that all lifestyle factors can potentially have a deleterious impact on fertility, especially obesity and smoking. Female obesity is reported to lower the chance of conception and increase the likelihood of polycystic ovary syndrome; even women with moderate obesity and an increased waist circumference are said to halve their spontaneous conception rate. In IVF the chance of pregnancy is reduced by around 30 per cent in obese women, and the risk of miscarriage increased by 30 per cent. Smoking has a similarly harmful effect on fertility in both spontaneous and assisted cycles. A study from the Netherlands compared the effect of smoking on IVF outcome to an increase in female age of 10 years—from age 20 to 30 years, and thus moving very close to the chronological age at which ovarian ageing begins to have a tangible effect.[9]

The issue at hand is where a reasonable level of indulgence ends and a deleterious excess begins. Its resolution is also complicated by the physician's added responsibility for the health of the child, over and above the usual parameters of patient treatment and cure. Thus, fertility doctors have a double responsibility—to the patient and the child—which is why guidelines may not always favour the couple.

Yet another complication in resolving this question of patient rights is the perennial question of how and by whom the treatment is paid for. Some form of restriction—whether by lifestyle factors or patient age—may only be considered reasonable when treatment is funded by government schemes. The question may seem different when treatment is fully paid for by the patient, who may well be prepared to accept a poorer outcome. Nevertheless, many clinics, working in the private or public sector, would not normally treat patients with a body mass index (BMI) above 35 kg/m²; and requests for treatment from women over the age of 42 would usually require general health reviews and counselling about the very poor outcome likely.

In the Netherlands, for example, a guideline drafted more than a decade ago set the age limit for IVF at 40 years. Some clinics still respect this limit, while many others (including ours in Utrecht) will accept older women in cases where screening for ovarian ageing suggests a favourable outcome. In women over 41, oocyte donation is usually recommended. However, the maximum age for any woman having any form of fertility treatment is set by the national medical association at 45 years. The vast majority of Dutch clinics still respect this recommendation—and the legal system will not intervene provided that the professional organisations adopt acceptable guidelines. In Utrecht all infertility patients are also seen in a 'periconception' clinic, where lifestyle factors (mainly smoking and body weight) are assessed in terms of a healthy pregnancy and a healthy child. Indeed, all our patients are encouraged to participate in an individually tailored lifestyle programme before starting their IVF, although no absolute criteria are set for starting treatment.

Similarly at the VUB in Brussels no decisions about fertility treatment are made without first examining the lifestyle habits of both the female and male partner. Smoking, drinking and weight are all important parameters of pregnancy outcome, but in our experience excess weight is the most negative predictor. Studies consistently show that ovulation is frequently disrupted in overweight patients—but can be restored to something like a normal cycle with no more than a 10 per cent weight loss. This may mean little more than a careful diet and a bit more exercise; while some will find this easy to achieve, others may need professional assistance. Our views are also emphatic that all our patients should stop smoking and excessive drinking before treatment begins.

The Infertility Epidemic

There is a catalogue of confusing and overlapping terms
to describe childlessness: the most common, 'infertility',
is sometimes referred to as 'subfertility', which more accurately reflects a reduced but not absent ability to have children.
Both, however, describe an inability to conceive within a fixed
period, usually 12 months. 'Sterility' is a more permanent and
absolute inability to conceive, while fecundity is the biological
ability to do so, often referred to by demographers in terms
of potential reproductive capacity. Almost all cases of infertility can now be treated by modern methods of reproductive
medicine, and most 'infertile' couples can be helped to have
children of their own.

But even at their best, people, unlike rabbits, are not efficient breeders. Around one in six couples will have difficulty

conceiving at some point in their reproductive lives and will meet the clinical definition of infertility. Even the fertile population has no better than a one-in-four chance of becoming pregnant when having regular unprotected sex during a natural menstrual cycle. And even in the most favourable circumstances, following unprotected sex just before ovulation, only one in three women will become pregnant. The rabbit, however, blessed with two uteruses and a gestation period of just eight weeks, can conceive while still pregnant and produce many litters during the course of a year. Some estimates suggest that a single pair can increase exponentially to almost 200 individuals in 18 months. In Australia, where rabbit infestation caused the closure of many farms, an initial release of 24 rabbits into the wild in 1859 had escalated to a feral population of more than 10 billion by 1926.

Humans are not so prolific, despite our overcrowded planet. Consensus from studies indicates a natural fertility rate of just 25 per cent per cycle, suggesting that pregnancy loss rather than pregnancy to term is the human norm. Why this should be is uncertain. The ovulated egg may not be fertilised in the Fallopian tube, the fertilised egg may fail to implant in the uterus, or the pregnancy itself may be lost before the next menstrual period. However, what is certain is that successful conception is not always easily achieved.

One study from Germany, which followed up more than 300 women trying to become pregnant, found that four out of five had become pregnant within six cycles; but after that, said the authors, 'serious subfertility must be assumed in every second couple' (that is, 10 per cent of the total).[1] Further analysis suggested around half of these 'subfertile' couples would in fact have 'a nearly zero chance of becoming spontaneously pregnant in the future'.

A review of several studies of infertility put prevalence in developed countries at between 5 and 15 per cent. This same study estimated a world infertility rate of 9 per cent. ICMART (the International Committee Monitoring Assisted Reproductive Technologies), which collects and co-ordinates the records of some 80 per cent of the world's IVF clinics, has recently reported that around 1.5 million IVF and ICSI cycles are now undertaken each year, reflecting a continuing annual growth in demand of around 10 per cent. Other calculations have suggested that this demand would increase much more if those who needed and wanted treatment had access to it. However, despite such limited access, ICMART has estimated, based on a 22 per cent delivery rate achieved in each cycle of treatment, that some 300,000 IVF and ICSI babies were born in 2006, with numbers still growing.

The 25 per cent pregnancy rate of natural conception is almost identical to the live birth rate achieved by IVF. The technology of assisted conception is thus able to achieve comparable results per cycle to nature, but—despite huge scientific effort—not yet improve on it. However, whether in natural or assisted conception, this near one-in-four success rate necessarily implies a three-in-four failure rate, and, for many couples embarking on a first cycle of IVF or just trying to get pregnant, disappointment is a likely outcome. And what lies behind this high rate of infertility are social trends and biological facts which we as doctors can do little about; for 'infertility' today is largely explained by factors related to the female partner's menstrual cycle and to her age.

'Natural' family planning methods, as formally developed by Drs John and Evelyn Billings, have been based on the very premise that the timing of sexual intercourse during the menstrual cycle has a make-or-break influence on the chance of pregnancy. Studies have consistently shown that conception is most likely during a six-day interval which ends with the presumed day of ovulation. A recent study of 140 fertile women, which assessed fertility each day with a hand-held hormone monitor, found that this 'six-day window' was in fact quite variable, with the most frequent phase lasting only three days, but many lasting more than six.

The key to the success of these natural methods of contraception is not just ovulation and the release of reproductive hormones, but also the build-up of a white mucus within the cervix which helps sperm cells pass freely through. A study conducted by the World Health Organization (WHO) in 1983 showed that the volume of this cervical mucus peaked on day 15 of the menstrual cycle; the WHO thus defined the fertile period as any day on which mucus was reported before the peak day until three days after the peak. In this study the average length of the fertile period was found to be 9.6 days, and the probability of pregnancy at its highest on the peak day.[2]

The essential ingredients for conception are an egg and a sperm cell, and the egg—of a healthy fertile woman—is delivered each month during ovulation. From puberty onwards, a miniscule 'leading' follicle within the ovary will begin to grow and mature during each menstrual cycle, to be released as an egg into the Fallopian tube. This usually occurs at the mid-point of each cycle after a 14-day period of growth regulated by hormones and known as the follicular phase. Once released into the Fallopian tube, the leading follicle released as an egg travels through the reproductive tract, hoping to meet a few sperm cells along the way. That's all as it should be. But real life shows that conception is rarely that simple, nor ovulation that regular.

Ovaries fit for purpose

The human female is almost (but not exclusively) alone in the animal kingdom to have a menopause. Most species, such as lions or baboons living in stratified social groups, tend to die soon after the birth of their final offspring. Like men, they remain fertile until they die. And in all species, including the human, the ovaries were designed to function throughout life. Indeed, it was only in recent centuries that average female lifespan began to outlast reproductive ability. At present, however, women can look forward to 30 or more years of active, purposeful life after their menopause—and even more years after the birth of their final child. For them, the menopause signals the end of their reproductive lives, but not of their chronological lives: cycles become irregular, and eventually ovulation and menstruation cease to occur, but life goes on. The ovaries have fulfilled their reproductive purpose, but women live on well into their eighties and nineties.

However, women were never intended to live to such an old age, and were never expected to have children so late in their lives. But, with better healthcare and nutrition, so the theory goes, women have outlived their allotted time frame, and so are destined to enjoy their lives in a state of sterility before death finally catches up with them. Moreover, as far as we know, the age of menopause has not much changed, even if chronological life expectancy has. Thus, the menopause remains anchored at an average age of around 51 years, while in developed countries life expectancy for women has now reached 87 years or so.

Of course, this extended life expectancy brings with it many disease conditions which seem mainly to affect women once they've passed the menopause: cardiovascular diseases, osteoporosis, loss of cognition, Alzheimer's, and urinary incontinence are all more common in postmenopausal women. In fact, the only major health benefit of an early menopause seems to be a reduced risk of breast cancer. But this aside, the ever increasing years which a woman lives after the menopause— and thus without the female hormone estrogen—represents a huge challenge for the future care of older women.

Thus, despite these challenges, in the human female at least, a reproductive life remains considerably shorter than a real life, and depends only on biology, not on evolutionary survival, medical breakthroughs or five fruit and vegetables a day. Indeed, the time span of a physiological reproductive life begins at birth—even if fertility cannot begin before puberty. Girls are born with a fixed supply of around 2 million follicles within their ovaries, and from that point on this supply will never increase. During each cycle, this stock of dormant follicles is depleted by around 1,000, and between 10 and 20 more developed follicles will begin to mature (encouraged by release of follicle stimulating hormone (FSH) from the pituitary). Of these follicles, only one will dominate to become the single egg dispatched at ovulation. Many other follicles will simply die over time, until, as the menopause approaches, the supply is nearing depletion and the capacity to reproduce is complete.

It is thus an indisputable fact that the strongest measure of ovarian reserve—that is, the capacity of the ovary to produce eggs viable for fertilisation and pregnancy—is age. The older the woman, the lower will be her ovarian reserve and her chances of conceiving. On average, a decline in ovarian reserve will begin from the age of 30, although with substantial variations among individuals.

In 1996 a benchmark study, which analysed every IVF treatment cycle performed in the UK between August 1991 and April 1994, cross-referred the live-birth outcome of each of those cycles with other factors characteristic of the individual patient. The results showed first that the overall live birth rate per cycle of treatment was 14 per cent, but the highest live birth rates were found in those women aged between 25 and 30 years.[3] Although younger women appeared to have somewhat lower rates, the steepest decline was in older women. The investigators also found that at all ages over 30 the use of donor eggs was associated with a significantly higher live birth rate than use of a woman's own eggs, suggesting that the age of the eggs, not of the woman, was the crucial factor.

Since then, the decline in fertility with age has been put into even sharper perspective—even on a year-by-year basis—by IVF data collectors in the US. In its report for 2006 the Centers

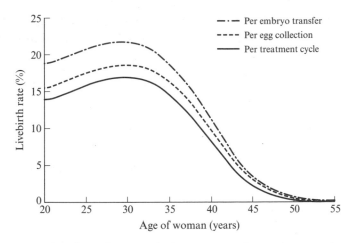

Figure 5 The effect of age on birth rate in IVF. The results were derived from the database of the HFEA and showed that the highest live birth rates were found in women aged between 25 and 30 years, with rates declining steeply after 35.

for Disease Control said unequivocally that 'a woman's age is the most important factor affecting the chances of a live birth when her own eggs are used'. CDC data for 2006 showed live birth rates falling from 30 per cent in patients aged 22, to 26 per cent at age 30, to 23 per cent at aged 35, to 12 per cent at age 40 and to just 1 per cent at age 45.

Live birth and fertility rates as a result of spontaneous conception also show similar declines with age, but the figures from natural pregnancies now add another complication to the story, with far-reaching implications for fertility clinics like ours. For what the latest population statistics show is a remarkable age-related change in the patterns of childbirth in the last few years. In the UK as recently as 1991 the greatest number of conceptions—281,500—occurred in the 25–29 age group, and the lowest—12,100—in the over 40s. By 2007, the 25–29 age group still had the highest number of conceptions, but total numbers had fallen dramatically to 234,200; the over-40s still had the lowest numbers, but their total had more than doubled to 26,500.

Analysing its 2007 data further, the UK's National Statistics Office confirmed that the highest percentage increases in rates of spontaneous childbirth were among older mothers. Indeed, in just one year—2006–2007—birth rates among women aged 35–39 rose from 54 live births per thousand to 57 per thousand in 2007, an annual increase of 5.7 per cent. Similarly, for women over 40 birth rate rose by 5.4 per cent between 2006 and 2007—from 11.4 to 12.0 live births per thousand. And in the US the number of first births per 1,000 women aged 35–39 increased by 36 per cent between 1991 and 2001, and in those aged 40–44 by a remarkable 70 per cent.

And all this despite an inevitable decline in natural fertility which appears to begin around the age of 35. So what's going on? How are birth rates increasing in the over-35s when their natural fertility is already on the wane?

What the European and US figures indisputably show is that women are choosing to have their children, or at least trying to conceive, at a later stage in their lives. The latest figures show that in many countries the average age of a women giving birth to her first child is now approaching 30. These are women putting off their first pregnancies for longer than they were even 20 years ago, and certainly for much longer than their grandparents did 50 years ago.

The result of delaying this first pregnancy attempt beyond the age of 30—and especially beyond 35—will inevitably lower the chance of success because of the adverse effect of age on ovarian reserve. The French demographer Henri Leridon has calculated that 75 per cent of women attempting a pregnancy at age 30 will have a conception ending in live birth within one year, 66 per cent at age 35, but only 44 per cent at age 40. And it's these age-related declines in fertility which are suddenly having huge repercussions in clinics throughout the world. For in fertility clinics too the average age of patients is also increasing year on year.

In the Netherlands, for example, the average age of women having IVF has now reached 38. Even more graphically, data from Britain's HFEA show that the average age of an IVF patient in 1992 was 33 years; but by 2007 this had risen to 36 years. Three years may not seem much in the grand scheme of things, but at this time in a woman's reproductive lifecycle, when her ovarian reserve may already be in serious decline, three years may well spell the difference between pregnancy and childlessness. For the fact is, despite the major advances we have seen in medical technology in the past two decades, there is still little that we can do to fully compensate for the natural decline in fertility which comes with increasing female age. IVF has little by little improved many of its processes and its success rates, and tried to meet this challenge, but 'making' a baby—any baby—is simply impossible for a woman who has exhausted her stock of ovarian follicles.

Investigating the infertile couple

The advancing age of our patient population has had a dramatic effect on fertility services in the past few years. Not only has the type of patient changed, but so has our approach. This is particularly so in our work-up to treatment and in diagnosing an explanation for our patients' infertility. Not too many years ago, text books decreed that the there were three causes of infertility split among couples in equal measure: one third of cases might be because of female cause (blocked tubes or hormonal problems, for example), one third of male cause (poor sperm count or morphology), and one third unknown (which may involve subtle abnormalities in both partners). Today, however, with more than half the total patients at some of Europe's main fertility centres over the age of 35, it's fair to say that the leading cause of infertility is simply related to female age and explained by nothing other than poor ovarian reserve. An accurate diagnosis may thus seem less relevant to many specialists than their patient's biological clock ticking away. Why wait months for an expensive surgical investigation to diagnose tubal adhesions in a 38-year-old patient, when the treatment will still be IVF?

Of course, work-up today still includes the basic diagnostic investigations which have long been used, but we are now adopting some tests more as a guide to the most appropriate treatment than to determine a precise cause of the infertility. These standard tests include semen analysis in the male partner, a visual assessment of the female partner's ovaries and reproductive tract (uterine and tubal structures), and a measure of ovarian reserve. The first two will indicate whether IVF or ICSI is the more appropriate treatment, while the third will determine how the patient is likely to respond to fertility drugs. With a greater role for these tests of ovarian reserve—and the increasing likelihood that we will never find a true cause of the infertility—there has been a shift in our work-up from a diagnostic to a more prognostic approach.

This means that in many cases we have given up on the basic medical paradigm of finding the true cause of an ailment. Instead, we take a pragmatic empirical approach which essentially depends on our assessment of a couple's chance of spontaneous or assisted pregnancy.

Semen analysis comprises a measure of sperm concentration, motility and morphology. However, there is a large degree of overlap between the sperm characteristics of fertile and infertile men, so definitive conclusions, except at extreme measures, are always difficult to draw. Similarly, prediction models based on sperm analysis alone should be interpreted cautiously. Most sperm tests are based on guidelines of the WHO, whose latest manual of 2010 defines 'normal' sperm as having concentrations of 15 million or more per millilitre of semen, a motility of at least 40 per cent, and the presence of at least 4 per cent of the sample looking 'normal' under the microscope.

However, evidence suggests that these WHO recommendations are not well observed in routine practice, with variable results. Nevertheless, semen analysis remains the principal diagnostic test of male infertility—and as an investigative test should be performed before all others. But in our view a greater standardisation of the tests and a more accurate laboratory evaluation are needed to improve their prognostic value. Semen analysis is useful, but not exclusively so.

The condition of the ovaries, uterus and Fallopian tubes can be visualised by a variety of techniques, including hysteroscopy, hysterosalpingography, transvaginal ultrasound, saline

Visualising the pelvic region

Hysteroscopy, a procedure used to examine the inside of the uterus via a thin microscopic camera (hysteroscope) passed through the vagina and cervix.

Hysterosalpingography provides x-ray images of the uterus and Fallopian tubes. A radiographic dye is injected into the uterine cavity through the vagina; the resulting image will show if the Fallopian tubes are open.

Transvaginal ultrasound provides magnified images of the cervix, uterus, ovaries and Fallopian tubes via a transducer passed through the vagina. High-frequency sound waves are returned to the transducer, and images are created from these echoes.

Saline infusion sonohysterography provides enhanced visualisation of the uterine cavity from trannsvaginal ultrasound for the detection of endometrial pathology; a saline solution is injected through the cervix as a contrast medium.

Laparoscopy is a minimally invasive (keyhole) surgical procedure which provides access to the pelvic region. The laparoscope allows surgical procedures to be performed and relays images back to a TV screen.

infusion sonohysterography, and laparoscopy. However, there is no consensus on which approach yields the best results.

Visualisation of the Fallopian tubes by ultrasound may reveal a fluid-filled blockage known as a hydrosalpinx. It was to circumvent such conditions that the original IVF procedures were devised by Edwards and Steptoe more than 30 years ago. There is still some controversy but a review of all studies on this subject concluded that laparoscopic removal ('salpingectomy') 'should be considered' for all women with hydrosalpinges prior to IVF. One of these studies suggested that the benefit of salpingectomy is greatest for patients with fluid-filled hydrosalpinges which are visible on ultrasound, and we agree that such patients should be encouraged to have prophylactic salpingectomy prior to IVF to improve their chances.

Uterine fibroids are detected in 20–50 per cent of women over 30 and are the most common benign tumour of the female genital tract. Small fibroids (4 cm or less in diameter) seem not to affect the outcome of IVF, while fibroids encroaching on the uterine cavity have an adverse impact. One study has found that removal of such fibroids can double pregnancy rates, but the evidence remains sparse. Similarly, the prophylactic removal of uterine polyps detected by hysteroscopy or ultrasound has also been shown to improve pregnancy rates in IVF and intrauterine insemination.

With both semen analysis and reproductive tract evaluation complicating the diagnostic model of IVF work-up, hopes have been building in recent years on the prognostic model promised by tests of ovarian reserve. Our centre in Utrecht has been at the forefront of research into these tests, particularly as a determinant of success—or failure—in spontaneous and assisted reproduction. Our aim is to find an accurate predictive model which will forecast the outcome of treatment in each individual patient, give us the information to counsel the patient about when to start treatment (or merely to wait and keep trying), and indicate which is the most appropriate form of treatment. So far, the prediction tests we are working on in IVF provide some measure of how a patient will respond to ovarian stimulation with gonadotrophins, but not on the likelihood of pregnancy and live birth.

Ovarian reserve primarily represents the number of follicles within the ovaries. In reality this number is most accurately assessed by measurement of hormone levels as a marker of ovarian reserve or by counting when visualised by ultrasound. Increasing evidence indicates that the number of developing follicles is representative of the stock of dormant follicles. However, the correlation between quantity and quality remains less clear. Indeed, it is still very difficult to estimate egg quality, but we do know from our experience in IVF that a poor response to ovarian stimulation (by which only a few follicles will develop in response to gonadotrophins) is often associated with poor embryo quality, low rates of embryo implantation in the uterus and an increased risk of miscarriage.

Not all women, however, react in the same way to ovarian stimulation in IVF, and not all respond normally. The poor responders produce fewer than four mature follicles, while others may over-respond, producing 20 or more follicles and running the risk of ovarian hyperstimulation syndrome. Poor response in IVF is an increasingly common challenge, brought about largely because of the older age of our patients and the impact of reduced ovarian reserve.

This present phenomenon is further complicated by the fact that ovarian reserve itself is so varied and unpredictable. Indeed, even in the healthy fertile population we see this variability in the different ages at which women reach the menopause, commonly ranging from 40 to 60 years. This means that some women will take 20 years longer than others to exhaust their follicle pool and deplete their ovarian reserve. Unfortunate young women with a premature menopause—or what is now becoming known as primary ovarian insufficiency—may even be at that point of no return before the age of 30, and for reasons which we still don't fully understand. So it is because of this individual variability in the extent of ovarian ageing that accurate tests of ovarian reserve have been developed, to predict response to hormonal stimulation and the outcome of treatment in terms of eggs collected and ongoing pregnancy.

The consequences of delaying a first pregnancy lie at the heart of today's infertility epidemic, and it's our view that such a challenge must be addressed first by public information on the age-related decline in fertility. But in the laboratory and clinic the challenge for the future will lie in our ability to assess the ovarian reserve of any woman at an early stage of her reproductive life. Right now, our models of measuring ovarian reserve are still on a 'damage done' basis; future research must turn to novel markers—such as subtle gene modifications—which can predict the rate and timing of follicle depletion.

So far, a measure of ovarian reserve has been made by two simple tests: a blood test to assess levels of a naturally occurring

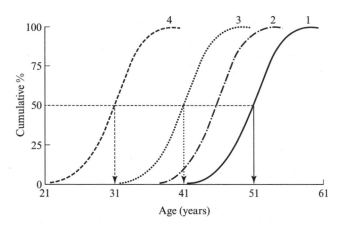

Figure 6 The variation in reproductive ageing. Curve 1 shows the percentage distribution of variation of age at the menopause, Curve 2 the variation of age of the transition from cycle regularity to irregularity, Curve 3 the variation in age of becoming sterile, and Curve 4 the variation in age of becoming infertile. Curve 4 indicates that infertility can begin at an age as young as 21, in young women experiencing premature ovarian insufficiency.

reproductive hormone known as follicle stimulating hormone (FSH); and a simple numerical count of early 'antral' follicles within the ovary (which appear at the beginning of the menstrual cycle) as seen by ultrasound. Studies suggest that antral follicle count correlates most closely with the number of oocytes retrieved at egg collection and is currently considered to be a good predictor of ovarian response to stimulation. In Utrecht we have also worked on another ovarian hormone—anti-Mullerian hormone (AMH)—and found that this too is an accurate predictor of excessive or poor response to hormone stimulation in IVF.

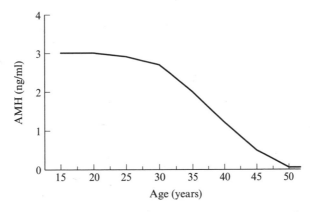

Figure 7 Levels of anti-Mullerian hormone (AMH) show a steady decline throughout reproductive life and are now known to reflect ovarian reserve and the decline in the follicle pool. Because of the strong correlation between AMH levels and follicle reserve, there are now indications that AMH measurement may be used to predict the time of the menopause.

AMH is a hormone expressed by cells within early follicles, with highest concentrations found in pre-antral and small antral follicles; AMH is thus thought to correlate closely with the number of antral follicles within the ovary. In Utrecht we have also found a good correlation between AMH levels in women aged between 30 and 35 and their likelihood of having a menopause some 10 years later. This increases the likelihood that decreased fertility (and low response to ovarian stimulation) is closely linked to age of menopause.

In recent years AMH has emerged as a strong predictor of ovarian response to stimulation, and thus as a measure of ovarian reserve likely to predict success or failure in IVF. A review of many studies did find—at least in the prediction of poor response to stimulation—that there was no difference between AMH levels and antral follicle count. However, the evidence for the predictive value of AMH in pregnancy (or non-pregnancy) was less robust. AMH, we reported from Utrecht, like antral follicle count, only represents the size of the cohort of follicles present in the ovaries, and the chance of pregnancy after IVF depends on many more factors than this cohort size alone.

So, in the absence of any other compelling evidence, it remains our view that age is the most important predictor of success or failure in IVF, and the reason why the developing world is now in the grip of the greatest epidemic of infertility ever known. If women delay their first attempt at pregnancy, particularly to a point beyond their 35th year, their chances of success will decline substantially. This is a simple fact which we can do little about. Indeed, researchers from Scotland, following a histological study of ovarian tissue, have recently proposed that 95 per cent of women by the age of 30 have only 12 per cent of their maximum pre-birth follicle population left in the ovaries—and by the age of 40 only 3 per cent remains. This is grim testimony to the fact that the female's store of eggs is defined at birth, and soon thereafter begins its inevitable decline to the menopause. Once gone, those eggs can never be replaced.

This inescapable fact is played out day after day in both natural and assisted conception, where female pregnancy and live birth rates each decline markedly after the age of 35. A calculation made by an ESHRE workshop on this subject found that the net rate of live births in a cohort of women seeking IVF treatment over a two-year period was 15.1 per cent for those aged 30, 9.6 per cent for those aged 35, and 0.9 per cent for those aged 40.[4] Similarly, the figures reported by the CDC in the US for non-donor IVF in 2005 show that singleton live birth rates for women aged 24 were 29 per cent but only 3 per cent at age 44. And in the age bands beyond 40, cumulative IVF data calculated from the HFEA database in Britain for the years 1992 to 2005 show similar trends—a live birth rate per started cycle of 8 per cent in the 40–44 age group but only 1 per cent in the 45–49 age group.

This decline in success rate is clearly seen in several randomised controlled trials performed at the VUB in Brussels. We find over and over again that a single embryo transferred in an IVF or ICSI cycle yields a delivery rate of almost 40 per cent in women under the age of 36, but in those aged between 38 and 43 the delivery rate falls dramatically to 17 per cent.

The depletion of eggs as women age is not the only explanation, nor the only function of ovarian reserve. There is also strong evidence that the quality of eggs—and thus of the embryos they form—also deteriorates as women grow older. This is most evidently seen in the number of eggs and embryos with chromosomal abnormalities, which appears to increase with age. These chromosomal abnormalities are now considered the main cause of embryonic wastage and pregnancy loss, and the main reason for the relatively poor pregnancy rates reported after IVF and natural conception, particularly in older women. In fact, an analysis of 196 embryos we performed in Utrecht found that only 36 per cent of the embryos were chromosomally normal three days after fertilisation. These were embryos from young patients, and other studies have shown higher rates of abnormality in older patients (even up to 90 per cent). Down's syndrome, as we have seen, a chromosomal abnormality in which there are three copies, not two, of the 21st chromosome (trisomy 21), is much more prevalent in older women than in their younger counterparts.

A landmark study from one of the pioneers of this work, Santiago Munné in the US, found in 1995 high rates of chromosomal aberration (aneuploidy) in both normally and abnormally developing embryos.[5] In a separate report that same year Munné concluded that aneuploidy rates increase proportionally with the patient's age.

Since then, the wealth of evidence we have on the chromosomal status of embryos suggests that herein lies the principal reason why embryos fail to implant in the uterus, or, having implanted, miscarry as a very early pregnancy loss. The explanation is unlikely to lie so persuasively in the uterus; results in the US from IVF cycles with donor eggs (always obtained from young fertile women) reflect high rates of implantation even in women over 40, and certainly much higher than those achieved with their own eggs. These high implantation rates seen in older women after the transfer of embryos derived from the eggs of a younger donor strongly suggest that the actual ability of the uterus to encourage implantation and maintain a pregnancy is largely unaffected by the recipient's age. What is affected by age and detrimental to the pregnancy is the quality of the embryo and the egg from which it is derived. And this seems most clearly determined by its chromosomal status.

However, while implantation and pregnancy seem as dependent on the quality of the original oocyte as on their number, studies have disappointingly failed to show any benefit in live birth rates when aneuploid embryos are identified and discarded before transfer. This has been our experience at the VUB in Brussels, where we showed for the first time that pregnancy rates in women whose embryos were screened for chromosomal disorders were no different from those who had no embryo screening. So far, the technology to detect the impact of these abnormalities remains elusive, but it nevertheless remains true that much of the promise of designing babies for tomorrow lies in these genetic screening techniques.

Why else do pregnancies fail?

While oocyte quality seems the principal explanation for the poor ongoing pregnancy rates following both natural and assisted conceptions, there are other reasons why pregnancies are not always achieved. The explanations can be attributed to three main causes: uterine dysfunction, ovarian dysfunction and sperm dysfunction.

First, despite the evidence provided by older women having egg donation, there may well be factors in the lining of the uterus (endometrium) which make it less receptive to the implantation of an embryo. Endometrial receptivity may be compromised by polyps and other pathological lesions, or by the consequences of ovarian stimulation itself. Some of these abnormalities may be visualised by hysteroscopy and confirmed by histology. Similarly, markers of a successful pregnancy—ovarian response to stimulation and implantation—may be compromised in patients with endometriosis, a common condition in which endometrial tissue develops outside the uterus. In Brussels we have also clearly shown that ovarian stimulation itself is associated with early secretory changes in the endometrium, which may make implantation less likely.

Recently, there have been hints of a breakthrough in detecting a receptive (or non-receptive) endometrium through the application of new molecular techniques—known as genomics—able to identify and quantify thousands of genes. Genomic studies have shown differences in endometrial gene expression during the different phases of the cycle, including the proliferative phase when the endometrium prepares to receive an embryo. But so far, these studies remain at the experimental stage, with little consistency in detecting which genes are indisputably associated with implantation.

Anyway, genes produce proteins, and it is proteins which appear to interact with the embryo on its journey to the uterus. This has been described as a 'cross-talk' in which the uterus, in its release of proteins, prepares the fertilised egg for implantation. There thus seems more promise in investigating these proteins as a marker of the endometrium's receptivity to implantation, rather than the specific expression of genes. The science of 'proteomics' has already identified several proteins which appear to be associated with implantation. And it may well be that these developments will provide a road map of the endometrium as it changes throughout the menstrual cycle. The patterns of protein expression which define the window of implantation may thus provide the blueprint for how an embryo implants in the endometrium and becomes a pregnancy.

There have been attempts to improve endometrial receptivity—and thereby pregnancy rates—by giving drugs which boost blood flow to the uterus and thereby increase the thickness of the endometrium (which needs to be around 9 millimetres to support a pregnancy). Both low-dose aspirin and sildenafil (Viagra) have been tried, the latter with rather more publicity than the former. A 2002 study at a Las Vegas fertility clinic found that vaginally administered sildenafil 'enhanced endometrial development' in 70 per cent of patients, and achieved 'high implantation and ongoing pregnancy rates'. Sildenafil, however, remains unlicensed for this purpose, and the evidence for its safety is non-existent.

Following failed implantation, the other main explanation for non-conception is ovarian dysfunction, typified most graphically in the very common condition of polycystic ovarian syndrome, or PCOS. As its name implies, PCOS is characterised by multiple cysts on the ovary, yet—nonsensical as it may seem—this is not a necessary feature in every diagnostic definition, especially in the US. PCOS is indeed associated with two other important diagnostic features: raised blood levels of male hormones (known as androgens), and irregular or absent periods. It is the latter characteristic which predisposes women with PCOS to infertility, but they are difficult cases because their multiple ovarian cysts raise risks from any reproductive treatment. A consensus meeting between European and American experts in 2007 concluded that the first-line medical treatment to help women with PCOS become pregnant was ovulation induction with a non-hormonal fertility drug known as clomiphene citrate. This, the experts said, would be the least likely to overstimulate the sensitive ovaries of PCOS and cause ovarian hyperstimulation syndrome and multiple gestations.

However, PCOS is also associated with high androgen levels (which may cause excessive facial and body hair) and in many cases excess body weight. The latter is in itself a risk factor for diabetes and cardiovascular disease and there are well-founded concerns that the 'phenotype' of PCOS first requires attention to these well-established risk factors for heart disease and stroke. But it is also well established that extremes of body weight— both over and under—have a disruptive effect on cycle regularity, and many overweight women with PCOS, while losing weight for the benefit of their hearts, will also find that their periods return to normal. Weight loss, therefore, is the first and safest fertility intervention for women with PCOS.

Indeed, all preconceptional care at our centre in Utrecht— whether for PCOS patients or not—is based on the well-founded belief that lifestyle does have an effect on fertility and on the health of parents and their babies. Our lifestyle assessment and advice concentrate on weight and nutrition, smoking and alcohol, and other underlying health problems. Diabetes, for example, is commonly associated with PCOS, and women with poor blood glucose control (that is, HbA1c levels of 10 per cent or more) are usually advised to avoid pregnancy altogether. Moreover, many of the medications prescribed to lower the risks of cardiovascular disease, such as beta blockers or ACE inhibitors, are not recommended in pregnancy. A thorough programme of preconceptional care will consider all these factors, and intervene with screening and structured advice where necessary.

Of course, there are other fertility treatments beyond weight loss and clomiphene in PCOS, but they all require the pre-administration of gonadotrophins to stimulate the ovaries; they thus carry the risks of overstimulation (OHSS) and multiple pregnancy. Monitoring the growth of follicles by ultrasound in PCOS patients given gonadotrophins for ovulation induction or IVF is essential to ensure that only one or two follicles are maturing and thus to prevent multiple pregnancies and OHSS.

Lifestyle changes in PCOS, however, may mean that fertility treatment of any sort is simply not necessary and conception will be readily achieved by patience and nature's way. These are couples who are genuinely 'subfertile', whose infertility is of a temporary nature and owes nothing to any physiological complication, and who are amenable to wait-and-see 'expectant management'.

What about the man?

Men, unlike women, do not have a menopause. Their reproductive ability seems independent of age and certainly does not come to an abrupt end with a final decline in sperm quantity and quality. Mid-life crises maybe, but no mid-life 'change' for them. Testosterone levels may decline with age, but characters like Charlie Chaplin and Luciano Pavarotti maintain the myth that the active male sperm cell is forever renewable.

Yet the truth is that age-related structural anomalies have been found in sperm cells, and men's fertility may well decline with age—even if with less emphatic biology than in women. Studies which found age-related changes in sperm count or morphology have so far been construed as clinically insignificant, but there is also evidence that the biological clock ticks for men as well as for women. For example, a recent follow-up study of more than 17,000 intrauterine insemination cycles in France suggests there may well be a marked age-related paternal effect on outcome. Investigators from Paris followed up 12,000 couples, measuring the husband's sperm count, motility and morphology before treatment. Clinical pregnancy, miscarriage and delivery rates were also recorded. As expected, the study found that advanced maternal age had a negative effect on the pregnancy rate and was also associated with an increased miscarriage rate. However, an exactly parallel effect was also found for paternal age, with pregnancy rates declining and miscarriage rates rising with the advancing age of the male partner. The investigators speculated that this might be because of changes in sperm DNA structure.[6]

The screening of sperm donors for sperm banking loosely recognises this possibility. In the UK, for example, the HFEA, which regulates sperm donation for fertility treatment, requires sperm donors to be between 18 and 45 years old. However, the latest guidelines produced by the British Andrology Society recommend an upper age limit of 40 for sperm donors, while adding a cautionary note that the number of mutations in the DNA of sperm increases with age.

Otherwise, the determinants of unhealthy non-fertile sperm are factors assessed in conventional sperm analysis— concentrations, motility, and shape. For the past 20 years there has been a consistent, if controversial, claim that sperm quantity and quality have each been falling because of environmental and occupational pollutants. A headline-grabbing study from the University of Copenhagen in 1992 found a 'genuine decline' in semen quality over the previous 50 years. The study had analysed a total of 61 reports published between 1938 and 1991, providing data on almost 15,000 men, and found that mean sperm count had fallen from 113 million/ml in 1940 to 66 million/ml in 1990. In addition, seminal volume was found to have declined from 3.4 ml to 2.8, indicating, said the authors, 'an even more pronounced decrease in sperm production than expressed by the decline in sperm density'. The investigators also noted a concomitant increase in 'genitourinary abnormalities' and speculated that some 'common prenatal influences could be responsible'.[7]

Meanwhile, what the Danes were suggesting in humans, others, especially in the US, were repeatedly reporting as fact in animals. Gender-bender changes in wildlife reproduction were described in a variety of species, raising such androgynous issues as feminisation, reduced fertility and altered sexual behaviour. Gastropods, reptiles, fish, birds and mammals had each been affected in some way or other in an alarming chronicle of wildlife catastrophes. The zoologists categorically laid the blame at the door of environmental pollution.

Since then, other studies have reported consistently declining sperm counts in humans. For example, in 1995 a French investigation found a 'true' decline over the previous 20 years, from 89 million/ml in 1973 to 60 million/ml.[8] The French investigators analysed the first ejaculate donated at a sperm

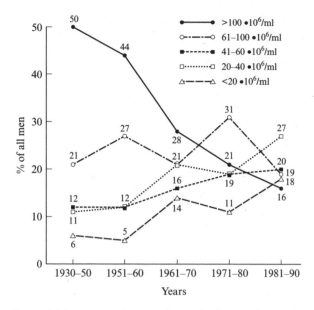

Figure 8 A 1992 report in the *British Medical Journal* caused a sensa‐tion by suggesting that sperm counts had almost halved over the previous 50 years. The most marked declines were found in sperm concentrations of 100 million per millilitre of semen, a value considered in the 'normal' range for fertility.

bank between 1973 and 1992 by each of 1,750 men. Results showed that the mean concentration of sperm decreased by 2.1 per cent per year. After adjustment for age and duration of sexual abstinence, each successive calendar year of birth accounted for 2.6 per cent of the annual decline in sperm con‐centration.

Not everyone accepts the environmental pollutant theory, and an editorial accompanying the French report in 1995 scoffed at the idea that men today are any less fertile than their brothers in history. 'Despite the declining sperm concentrations reported...there is little evidence that male fertility is declining,' wrote the editors, but the evidence suggests to us that there is indeed a universal fall in sperm count and morphology which is not easily explained. Right now, it is not possible to say what effect this is having on male fertility, but at the VUB in Brussels, where we have many male factor cases referred for treatment by ICSI, we are not aware of any obvious trend in declining sperm counts. Of course, our cases are by definition not representative of the general population, but at least even the most severe cases of azoospermia and oligozoospermia can be successfully treated (provided the female partner's age is acceptable).

There is evidence, however, that occupational pollutants—such as pesticides and lead—and lifestyles can and do have an adverse effect. Smoking, excess alcohol and drug abuse have all been linked to reduced male fertility, though the evidence is not consistently strong. For example, a small recent study from Italy found that sperm concentrations in heavy smokers were significantly lower than those observed in mild and non-smokers, although some parameters of semen quality (in men with unexplained infertility) were not 'dramatically affected' by cigarette consumption. However, in a study of 301 couples treated at a German fertility clinic the pregnancy rate following ICSI in women with smoking male partners was 22 per cent but 38 per cent in those with non-smoking partners. Smoking by males, concluded the investigators, decreases the success rates of assisted reproduction procedures. And we too agree that male smokers (as well as female) considering natural or assisted conception should quit as an essential requirement of their preconceptional care.

The demographics of infertility

In July 2007 the Spanish government announced that the family of every child born in Spain would receive a cash donation of €2,500. This baby bonus, explained prime minister Jose Luis Rodriguez Zapatero in his state-of-the-nation address, would be Spain's bid to help raise the country's languishing birth rate and support its fast growing economy. 'Spain needs more families with more children,' he said, 'and families need more support to have these children.'

Spain's fertility rate—of 1.37 children per woman of childbearing age in 2006—was and still is among the lowest in Europe. But Spain is certainly not alone. Figures from the EU show that almost all countries in Mediterranean and Eastern Europe now have fertility rates far below the 2.1 children per woman thought necessary for population replacement. Indeed, most of the lowest fertility countries of the world are now in these two areas of Europe.

With a pinch of understatement, a white paper communication produced for the European Commission in 2006 described Europe's demographic decline as a 'challenge', not a crisis. The overall fertility rate was put at 1.5 children per woman, with no more than a 0.1 increase forecast by the year 2030. The consequences, warned the report, would be a 'spectacular' increase in the number of old people needing social and financial support. The problem would be compounded by a continuing rise in life expectancy, with a dramatic increase in people in their eighties and nineties. Its impact would reverberate into the labour market and economic growth, into social security and public finance.

The Commission concluded that an 'overall strategy' would be needed to address the social barriers to having children (more educational and child-care services for parents, better managed working hours, extended pregnancy leave, tax concessions) but it also singled out 'the substantial impact of the rise in the age at which women have their first child, reflecting the growing reluctance of couples to have children'.

This growing reluctance to have children, or at least the trend to defer the birth of a first child, has been apparent in our fertility clinics since the early 1990s, with around one in three of all IVF cycles now performed in women aged 35–39, and more than one in 10 in women over 40. Thus, what we seem to be seeing as demand for IVF continues to grow and fertility rates remain at their lowest, is a fertility epidemic determined principally (though not exclusively) by the deferment of a first pregnancy attempt, a medical complication related more to age than to any disease or physiological problem. In fact, a 'research note' for the European Commission reported that a one-year rise in the average age of the mother at first birth coincides with a 5 per cent increase in childlessness within her same birth cohort. So it seems indisputable, as the European Commission's white paper suggested, to assume that this trend in delayed childbirth is indeed reflected in Europe's falling fertility rates and in the ever increasing numbers in our fertility clinics.

The decline in Europe's fertility rates 1960–2004–2008, expressed as the mean number of children born to each woman per country

	1960	2004	2008
EU-25	2.59	1.49	
Austria	2.69	1.42	1.41
Belgium	2.56	1.64	1.82
Czech Republic	2.11	1.22	1.50
Denmark	2.57	1.78	1.89
Estonia	–	1.40	1.66
Finland	2.72	1.80	1.85
France	2.73	1.90	2.02
Germany	2.37	1.36	1.37
Greece	2.28	1.29	1.45
Hungary	2.02	1.28	1.35
Ireland	3.76	1.99	2.03
Italy	2.41	1.33	1.41
Latvia	–	1.24	1.45
Lithuania	2.60	1.26	1.47
Netherlands	3.12	1.73	1.77
Poland	2.98	1.23	1.23
Portugal	3.10	1.40	1.37
Slovenia	3.07	1.25	1.46
Slovakia	3.07	1.24	1.33
Spain	2.86	1.32	1.46
Sweden	2.20	1.75	1.91
UK	2.72	1.74	1.94

*Data 1960/2004 from Population statistics, European Communities 2006
*Data 2008 from Eurostat, Population and social conditions, Data in focus 31/2009
<http://epp.eurostat.ec.europa.eu/cache/ITY_OFFPUB/KS-QA-09-031/EN/
KS-QA-09-031-EN.PDF>

The facts are simple and beyond dispute: women are putting off the time at which they want to start their families. And this now seems a social phenomenon almost beyond the reach of any public health campaign. Its cornerstones are found in the greater availability of higher education (the link between educational attainment and delayed pregnancy is unequivocal), a greater emphasis on career and consumption, reduced relationship stability, and more economic uncertainty. These are trends with deep social roots, whose advance will be little influenced by demographers or well-meaning doctors like ourselves. For what we are now seeing in our fertility clinics are the consequences of these social trends: women in their mid-30s who suddenly find that their fertility is not as buoyant as anticipated, and who, without much warning, have become yet another statistic in Europe's growing infertility epidemic. And, as our tests of ovarian reserve indicate, there is often very little we can do to help.

The Right Treatment for the Right Patient

Before we decide on the most appropriate form of treatment for a couple, some degree of fertility work-up will have been completed in the form of semen analysis, hormone measurements for ovulatory dysfunction and some visual investigation for anatomical problems in the pelvic region. A precise history-taking will also be important for an idea of cycle regularity and past fertility. These preliminary steps might even indicate that treatment is not the best option at this time, and expectant management for a year or so a better alternative.

Indeed, one of the challenges of modern IVF is determining whether expectant or active management is the more appropriate for each individual couple. For the fact is that many 'infertile' women will become pregnant while waiting for treatment; some patients even become spontaneously pregnant after unsuccessful IVF. The difficulty is in deciding who they are, and such decisions are helped by a predictive assessment in which several factors—not one—all contribute to the outlook. One study from the 1990s, which followed up more than 2,000 infertile couples, found that 14 per cent of them became pregnant before treatment started and within a year. The main predictive factors were pregnancy history, duration of infertility, female partner's age, a male defect, endometriosis, and tubal disease.

Because any reproductive treatment is stressful and exposes the patient to medication and some risk, the decision to treat actively and not expectantly should be carefully considered and based on the prospects of each individual couple. Overtreatment is a potential source of harm. However, this principle is less easy to apply in a female patient over the age of 35 whose biological clock is rapidly ticking away. Thus, while the decision to treat should not be based solely on a simple definition of 'infertility', expectant management becomes less appropriate with advancing female age.

Once a decision to treat has been made, the options lie with the assisted reproductive technologies of IVF and ICSI, egg donation, ovulation induction in women with absent cycles (anovulation), insemination with donor or partner sperm, or reproductive surgery.

With such techniques available there are very few couples with infertility who cannot now be helped. Even young women with premature ovarian failure can be helped if they are prepared to accept a donor's egg. Today, therefore, the emphasis in treating infertility is to apply the most appropriate treatment to each individual couple.

It's rare these days that our fertility work-up unravels a true cause of infertility, and IVF often represents a reasonable treatment regardless of cause. Indeed, IVF is usually an empirical intervention in most of those who are treated with it. And, as we have seen, the classical medical paradigm of history-taking, investigation, diagnosis and treatment has proved of declining value in the modern management of infertility. We should, however, emphasise that a proper work-up to treatment is still valid and important—because some diagnosed causes of infertility can be treated successfully, and sometimes other diseases—such as diabetes, thyroid disease, or even pituitary tumours—can be identified, and these require their own dedicated treatment. A proper work-up will also be important for the couple to know if they still have a reasonable chance of spontaneous pregnancy.

In vitro fertilisation

While it is convenient to describe all assisted reproductive treatments as 'IVF', such an acronym does little to reflect the complexities of these treatments today. As we have already seen, the traditional procedure of IVF as pioneered by Edwards and Steptoe in 1978 was a basic approach whose application was and remains the classical indication for treatment: to bypass a tubal blockage in a female patient. In those same patients today—those with hydrosalpinges, tubal adhesions and scarring of the Fallopian tubes—traditional IVF remains the most appropriate treatment. If a tubal obstruction prevents the passage of an ovulated egg from the ovary to the uterus and along the way its rendezvous with a sperm cell, fertilisation can be achieved in the laboratory, if not in nature.

Although patients with tubal problems today represent only a small proportion of those having IVF, it is from this original basis that the multiple techniques of assisted reproduction are derived. Not until the introduction of ICSI, however, in the early 1990s could severe male factor infertility be treated by an effective IVF method. Until then, there was little that conventional IVF could do to compensate for sperm samples of poor quality or meagre quantity. And not until the techniques of surgical sperm retrieval from the testis or epididymis were developed could even ICSI help men with absent sperm in their ejaculate.

Most forms of IVF today (though not all) require a drug phase of pre-treatment to 'hyperstimulate' the ovaries to produce multiple follicles. In natural conception ovulation is triggered by a surge in luteinising hormone (LH), a reproductive hormone produced in pulses by the pituitary gland at mid-cycle. In assisted conception this same reaction is triggered by an injection of human chorionic gonadotrophin (hCG), which also triggers oocyte maturation. Egg retrieval, through the aspiration of follicular fluid, takes place 36 hours later, just before spontaneous ovulation would have occurred. However, should a premature surge of LH occur before the administration of hCG, the multiple follicular growth encouraged by the ovarian stimulation would be compromised and the treatment cycle cancelled.

This potential problem in IVF was solved in 1982 when a group from Glasgow reported that five infertility patients were rendered 'hypogonadotrophic' with large doses of an agonist of LH-releasing hormone. Thus, what this agonist did was first stimulate and then shut down the natural hormonal cycle before ovarian stimulation for IVF actually began. This avoided the risk of a premature surge of LH and treatment cancellation. Within two years a substantial report in the *Lancet* showed that a GnRH agonist—used in combination with gonadotrophins for ovarian stimulation—reduced cancellation rates, improved IVF outcome and allowed greater flexibility in the timing of hCG and egg retrieval.

It is this protocol—long pituitary downregulation with a GnRH agonist and ovarian stimulation with gonadotrophins—which has defined the drug treatment of IVF for the past 25 years. The 'long agonist protocol' was universally welcomed by fertility clinics for its low cancellation rate, high numbers of ovulatory follicles (as well as eggs and embryos), and improved pregnancy rates. However, in time there were also others who reported that this long agonist/high dose protocol was time-consuming, expensive and disagreeable to patients. The long protocol had potential for discomfort, and high drop-out rates.

The combination of a long agonist protocol with a high starting dose of gonadotrophins has also been associated with an increased risk of ovarian hyperstimulation syndrome. A 2003 review of OHSS described it as a 'rare' complication of ovarian stimulation, developing in its early form a few days after the follicles have been punctured to release their eggs. Symptoms are usually worse (and can even be fatal) at later stages. OHSS symptoms range from abdominal distention and discomfort in mild cases to severe enlargement of the ovaries. Studies of adverse events associated with IVF have put the incidence of mild OHSS at around 5 per cent, and severe reactions at around 1–2 per cent. A recent report has suggested that a 'key element' in the cause of the syndrome may be a genetic change in the FSH receptor; however, other reports have implicated many other possibilities, ranging from an imbalance in the level of electrolytes to an alteration in hormonal dynamics. What is clear, however, is that OHSS is most severe in the presence of pregnancy, when the hCG produced during early pregnancy stimulates the symptoms.

The risk of OHSS in the long agonist/high dose protocol has been reduced by the more recently introduced mild approaches to IVF, reflected in lower doses of gonadotrophins and the transfer of just one embryo (and therefore a need for fewer eggs). Our work in Utrecht has been at the forefront of these approaches. However, mild stimulation does not entirely remove the risk of OHSS and cycle cancellation, suggesting that 'mild' may not be mild enough for some patients.

The long agonist protocol was what it said—a long treatment. This is because the GnRH agonist—administered, for example, as buserelin by nasal spray or leuprolide as an injection—does not immediately inhibit the release of FSH and LH from the pituitary, but takes around two weeks to 'downregulate' pituitary activity. Thus, a full course of IVF treatment using a long agonist protocol will need around four weeks from the first administration of the agonist to the injection of hCG and egg collection.

By contrast, a GnRH antagonist has an immediate effect on the pituitary and can suppress the natural release of FSH and LH within a few hours of injection. This means that a GnRH antagonist can achieve the same effect as an agonist, but without the initial flare of hormones and long period of downregulation. In contrast, antagonists work immediately, allowing use to be limited to just a few days during stimulation. And because they do work immediately, an antagonist—such as cetrorelix or ganirelix—injected after the start of ovarian stimulation can reduce total treatment time from around four weeks to just two or less. This is a real benefit for the patient, whose anxieties about treatment will inevitably be reduced.

But the benefits of an antagonist may be much more than just a shorter treatment time for the patient. The long protocol with an agonist, because it brings to an end the body's own reproductive hormone cycle, in effect causes a brief menopause, and with it the classical symptoms of estrogen deficiency which occur at this time—hot flushes, night sweats, vaginal dryness, mood swings. However, treatment with a GnRH antagonist is not associated with estrogen deficiency, nor with menopausal symptoms. Studies show that patients treated with the long GnRH agonist protocol report significantly more physical discomfort during the week before ovarian stimulation than those given an antagonist.

But the greatest advantage in our view is a lower risk of OHSS in patients treated with antagonists. OHSS risk can be

predicted by screening for risk factors such as polycystic ovaries, high estrogen levels or a relatively high number of small
follicles during stimulation. In such circumstances we have traditionally had two possible ways to minimise the OHSS risk:
first by stopping the course of gonadotrophins and allowing
the treatment cycle to 'coast' forward; and then by collecting
the available eggs for embryo freezing and transfer in a later
cycle. However, it is now clear—as a review from the evidence-
based Cochrane library has confirmed—that GnRH antagonist
therapy in IVF is also associated with a substantially lower risk
of severe OHSS than pretreatment with an agonist.

Co-treatment with a GnRH antagonist allows further reduction of the OHSS risk by replacing the single shot of hCG to
trigger final egg maturation; hCG at the end of the stimulation
phase is invariably the catalyst for OHSS. However, in antagonist cycles egg maturation can also be achieved with a single
injection of GnRH agonist, thereby avoiding the use of hCG
and reducing even further the risk of OHSS.

In our view the use of GnRH antagonists represents a substantial benefit to patients and clinics: a shorter treatment time
which reduces patient anxiety and distress, a lower exposure
to total drug dose (and thus lower cost), and a marked reduction in the first and foremost treatment risk from IVF.

So why are GnRH antagonists still neglected in many IVF
treatment programmes, in favour of the more cumbersome and
protracted long agonist protocol? We must assume that many
of these clinics are comfortable with the protocols they have
always used. Control of the cycle with an agonist also made
timed scheduling simpler, with no need for egg collections at

the weekend. But there is also a likelihood that many clinics are still concerned about success rates following the early clinical trials of agonists and antagonists, which found slightly higher pregnancy rates in patients given the agonists. This is an ill-founded anxiety, for the most recent studies and reviews have found no difference in outcome between the two. Indeed, the most recent trials of GnRH antagonists have reported higher pregnancy rates than earlier trials, which used protocols that would now be recognised as less than optimal.

It's our opinion that the GnRH antagonists should now be included in a package of individualised IVF approaches which also includes single embryo transfer, low dose ovarian stimulation, the freezing of supernumerary embryos (and their transfer in subsequent cycles), and the assessment of 'success' in terms of cumulative healthy live births per treatment started.

Such parameters defined the 'mild' IVF model which we tested in Utrecht in what was the world's largest randomised trial performed without drug industry support. In this 12-month study, 404 IVF patients were randomised to either the mild protocol with GnRH antagonist and single embryo transfer or a more conventional protocol of standard ovarian stimulation, GnRH agonist and two-embryo transfer. The shorter treatment time with the antagonist allowed more IVF cycles over the treatment period, had fewer drop-outs and was cheaper. The mild approach led to comparable live birth rates over the full 12-month treatment period (44 per cent for the mild strategy and 45 per cent for the conventional). Moreover—and not surprisingly—the proportion of couples with multiple pregnancies was just 0.5 per cent with the mild IVF treatment but 13.1 per cent with standard treatment. It was in planning the study that we proposed a change of emphasis in defining 'success' away from the mantra of pregnancy rate per cycle and more to the concept of cumulative, singleton live birth per initiated treatment.

Of course, there will be some infertility patients—those with a predicted poor response, for example—for whom this mild IVF package may not be suitable. But for the majority of normal responders we believe it will provide their best possible treatment experience, with minimal discomfort, minimal risk, and minimal cost. Moreover, if success is measured according to cumulative outcomes, results will not be compromised. Despite our promising initial results, there is even more scope for improvement if the ovarian stimulation phase of mild IVF can be fine-tuned according to individual patient characteristics.

However, for the main part progress in IVF has been driven by developments whose sole objective is to improve live birth rate—both to give patients a better chance of success and improve the clinic's own performance profile. Thus, applying 'the right treatment to the right patient' is not just a question of minimising risk; it's also a matter of maximising the prospect of success. And the various registries of IVF show that performance is indeed improving. For example, the records of the HFEA in Britain show that on average 23.7 per cent of all IVF and ICSI treatments in 2007 resulted in a live birth, a modest but real increase of 0.6 per cent over the previous year.

When reported in the medical journals, some of these attempts to improve pregnancy and birth rates seem impressive, with success rates of 70 per cent or more recorded. But invariably these results are in small studies with highly selected patients who rarely reflect the spectrum of couples seen every day in clinics. Moreover, many of the studies reporting success rates over 50 per cent are dependent on no more than a positive pregnancy test per embryo transfer. This is a long way from a live birth per started cycle, let alone a full course of treatment.

Over the years there have been many developments designed to improve outcome. Some aim to improve fertilisation, others sperm quality and quantity, uterine receptivity, or more precise embryo selection. One of these developments now gaining worldwide interest is the extended culture of embryos beyond the usual two or three days before transfer, for up to five days when the embryo reaches the blastocyst stage. In many clinics blastocyst transfer is now becoming a reasonably preferred option over conventional day two or three embryo transfer.

In 1998 embryologists working in Colorado shocked the world of IVF by reporting a remarkable 71 per cent pregnancy rate following blastocyst transfer in a small study group of patients. There were many who simply didn't believe what they read, claiming in a heated journal correspondence that the results were more a reflection of careful patient selection than the transfer of blastocysts. But, as the Colorado group insisted, this was a randomised trial, and the control group, who had day two or day three embryo transfers, were drawn from the same pool of patients as the blastocyst transfers. The

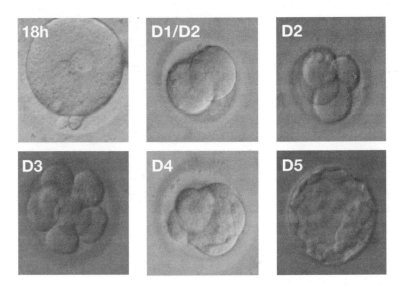

Figure 9 Embryonic development showing morphological changes at 18 hours, day one, day two, day three (cleavage stage), day four and day five (blastocyst stage). Recent trials suggest that transfer at the blastocyst stage is associated with higher implantation rates than at the cleavage stage.

outcome, they argued, could only be because of an effect of the transfer technique.[1]

Even then, the concept (if not the reality) of transferring blastocysts and not embryos in IVF was nothing new. The argument was that in nature the fertilised egg takes around five days to travel through the reproductive tract to the uterus, and what implants in the endometrium is not a two or three-day-old ('cleavage stage') embryo. Rather, it is an ever expanding blastocyst which is now beginning to break out of its embryonic shell. Blastocyst transfer, therefore, was always deemed a more physiological technique than cleavage-stage transfer.

However, the appeal of blastocyst transfer lay not just in apparently higher pregnancy rates, but also in the higher rates of implantation seen in most studies. And it was this which made blastocyst transfer the favoured technique for single embryo transfer. One blastocyst would have a higher chance of implanting than a day three embryo, and thus a more likely chance of success. This likelihood was recognised in 2007 by a Cochrane review which found 'evidence of a significant difference in pregnancy and live birth rates in favour of blastocyst transfer…[and] that in selected [good prognosis] patients blastocyst culture may be applicable for single embryo transfer'. This too has been borne out by studies, with most blastocyst transfer cycles showing a reduction in multiple pregnancy rate. For example, an analysis of 2,451 IVF cycles performed at Guy's Hospital in London found that the pregnancy rates in the single blastocyst transfer cycles were not only higher than the cleavage-stage transfers, but that multiple pregnancy rates were cut from 32 per cent to 17 per cent. This study underpinned the HFEA's commitment to reducing UK multiple pregnancy rate to 10 per cent.[2]

The American Society for Reproductive Medicine (ASRM), which sets guidelines for assisted reproduction in the absence of legislation in the US, updated its recommendations on embryo transfer in 2009 such that for patients under the age of 35 with a 'favourable' outlook a limit of one blastocyst (the equivalent of one or two cleavage-stage embryos) was recommended. And in the UK, the expert group advising the HFEA reported: 'Where more than one good quality blastocyst is available for transfer on day 5 or 6 of culture, the case for single blastocyst transfer is overwhelming.' Indeed, patient information supplied by the Society for Assisted Reproductive Technologies (SART), the national IVF registry in the US, has described the risk of multiple pregnancy as the principal indication for blastocyst transfer.

With such powerful support behind it, blastocyst transfer should be the convention in IVF, but it is not. There are still problems, particularly in identifying at day two or three which embryos are likely to benefit from extended culture and become healthy viable blastocysts. Some embryos lose their shape and integrity, others simply arrest in their growth. And so some women—who may have had four or five good quality embryos at day three—suddenly have no (or very few) good quality blastocysts at day five. For them, having gone through such a stressful programme of ovarian stimulation, egg collection and fertilisation, there will be no embryo transfer and no chance of pregnancy in this cycle, and no embryos for freezing for a further cycle. Past studies have suggested that as many as 40 per cent of patients taking the blastocyst route might have nothing to transfer at day five. And of course, as with cleavage-stage embryos, not every blastocyst transferred will implant as a pregnancy. Moreover, some which failed to implant as a blastocyst may well have implanted as a day two or three embryo. In addition, research is still needed into what effects the extended culture might have on offspring; in cattle breeding, extended culture has not been without developmental problems.

Nevertheless, as we will see in the next chapter, some of the greatest recent developments in IVF have indeed been seen in the IVF lab, not in the IVF clinic, and progress now in the application of culture media, incubator conditions (including air quality) and standardised quality control make the failure of blastocyst transfer less and less likely. A strong culture environment in a well-controlled laboratory system will mean that the embryo is under minimal stress, and less likely to be compromised by the extended time in the incubator. Blastocyst transfer only works well in clinics with top class IVF laboratories.

But it now seems clear that an embryo which does grow strongly into a blastocyst does demonstrate its own competence, and thus has an excellent chance of becoming a pregnancy. For example, a randomised trial we performed at the VUB in Brussels in 351 women under the age of 36 found a far higher delivery rate following blastocyst transfer (32 per cent) than cleavage-stage transfer (23 per cent). In fact, the trial was stopped early after interim analysis because of the difference in outcome. Similarly, a recent ASRM report calculated a live birth rate of 29 per cent with cleavage-stage transfers and 36 per cent with blastocysts in selected good prognosis patients; however, in less favourable patients (for example, of an older maternal age or with at least one failed IVF cycle behind them), live birth rates were described as 'not significantly different' between day three and day five transfers.

Selection of the perfect embryo guaranteed to implant and become a healthy singleton baby has been described as the holy grail of IVF. Blastocyst transfer is not yet that holy grail. Certainly, extension of the laboratory culture period to day five appears to improve implantation rate, and with it the greater opportunity for single embryo transfer and higher live birth rates in some patients. But, despite the speed at which technology is now advancing in the IVF laboratory, we are still not sure how best to select those embryos most likely to become viable blastocysts. We still need greater standardisation of blastocyst morphology and objective measurements of growth kinetics throughout the five-day culture period to increase the validity of our results.

Moreover, because of the faster preimplantation development of male embryos, there have been fears that blastocyst transfer favours the birth of boys over girls. In one retrospective study almost 58 per cent of babies born as a result of blastocyst transfers were male—in contrast to 51 per cent in the general population. There were also fears—later discounted—that blastocyst transfer was associated with a slightly higher rate of monozygotic twins (identical twins resulting from a split embryo), although a subsequent report has described the monozygotic twin rate in blastocyst cycles as 'low' and no cause for concern.

Thus, although blastocyst transfer might be a step in the embryologist's quest for the holy grail, it is certainly not the end of the road. The viability—even suitability—of an embryo depends on much more than its competence to form a blastocyst, and it's in the selection of that embryo that the real excitement in IVF now lies.

Egg donation

Egg donation is a development of IVF in which the eggs are provided by a donor, not by the patient herself. Egg donation is thus indicated in women unable to produce their own eggs, usually because of premature or age-related ovarian failure. Other indications have included diminished ovarian reserve, infertility following treatment of early onset cancers, and repeated IVF failure. Thus, eggs obtained from a suitable donor (either known or unknown) are fertilised with sperm from the recipient's partner and the resulting embryos are transferred into the recipient's uterus. The treatment has become relatively common today, with both demand and success rates reportedly high.

The key to that success depends on synchronisation between the short period of uterine receptivity in the recipient patient and the readiness of embryos for transfer from the donor. This synchronicity is usually achieved by the administration of hormone replacement therapy (HRT) in the recipient. The lining of the uterus is primed with HRT to mimic the normal fluctuations of ovarian hormones in a natural cycle in order to create an endometrial thickness (at least 8 mm) and hormonal environment consistent with implantation. Thus, what distinguishes egg donation from conventional IVF is that the process of ovarian stimulation in the donor (to provide an adequate pool of oocytes) is completely separate from the preparation of the uterus in the recipient for embryo transfer.

Even though the first pregnancy from a donor oocyte was reported almost 30 years ago—from the pioneering group of Carl Wood and Alan Trounson in Melbourne—the technique is still highly controversial today. Issues under discussion include payment to donors, the supply of donor oocytes, anonymity, as a treatment for older postmenopausal women, and as a major reason for 'fertility tourism'. In addition, there are several jurisdictions and cultures which do not accept the involvement of a third party in fertility treatment. For example, egg donation is outlawed in Germany and Austria, even though the ban in Austria has recently been challenged under European human rights legislation.

The original egg donation treatment reported from Melbourne in 1984 was in a young woman with premature ovarian failure (a premature menopause, in fact). And this remains an important indication for egg donation. The first symptom is usually an absence of regular menstrual periods in women under 40, and the diagnosis is confirmed by detection of raised FSH and low estrogen concentrations in the serum, suggesting a primary ovarian defect. Such symptoms are invariably a sign of infertility, which can have devastating consequences when it occurs at a young age. The cause of primary ovarian insufficiency (POI) is not well understood, but certainly involves mutations in certain genes known to be involved in reproductive function; the outcome is always a premature exhaustion of the resting pool of primordial follicles and depletion of ovarian reserve.

Management of POI in young women is not just about fertility. Diagnosis can have a deep psychological effect, and estrogen deficiency can have both short-term (hot flushes) and long-term (osteoporosis) consequences. The latter can usually be prevented by hormone therapy.

However, today it seems likely that egg donation is most commonly indicated not in POI but in the treatment of older women, particularly those over the age of 40—and even, as we shall see in chapter 7, in postmenopausal women over the age of 50. Certainly, egg donation as a treatment for infertility is on the increase, even in countries where the supply of donor eggs is limited. In the US, where there are no regulatory restrictions on payment to donors, the number of egg donation cycles increased to 14,000 in 2007, representing 11 per cent of all IVF treatments. In Europe, this proportion is much lower. Data from ESHRE's IVF monitoring showed that 13,000 egg donation cycles were performed in Europe in 2006, a small increase on previous years, but still only about 3 per cent of the total cycles. Success rates are also increasing. In 1997 ESHRE data showed a pregnancy rate of 27 per cent per transfer; by 2006 this had increased to 44 per cent. Even more striking, when stratified for age, pregnancy rates per cycle actually increased with the age of the recipient—from 37 per cent in those aged under 34 years, to 40 per cent in the over-40s. This latter trend reflects what has long been known in egg donation—that a successful outcome depends not on the age of the oocyte recipient, but on the age of the donor.

The uptake—and availability—of egg donation seems directly related to the availability of donors, and this in turn seems related to payment. A 2009 case report of fatal colon cancer in a young, previously healthy woman four years after repeated ovarian stimulation for egg donation suggested that more than 100,000 non-patient women in the US had donated oocytes for the treatment of others. Large sums in payment to suitably screened donors have been reported, but in most countries of Europe, where such commercialisation of compensation is not allowed, the pool of oocyte donors is considerably less. Nevertheless, payment to donors remains an issue of great controversy, with some arguments citing a higher rate of compensation as the solution to a chronic shortage of donor oocytes, while others insist that donation should be inspired principally by altruism. In Europe, many oocyte donors for treating cases of POI are indeed from friends and family. What is certain is that oocyte donation is not a simple risk-free undertaking for the donor, who must undergo screening and ovarian stimulation to provide her supply of eggs.

Ovulation induction

Ovulation induction aims to stimulate the development of a single dominant follicle in anovulatory women through the administration of fertility drugs, which were first used more than 50 years ago.

Advances in our understanding of the dynamics of ovarian physiology led to the discovery of gonadotrophic hormones, and they were introduced in the management of infertility from the early 1930s, when the first pregnancies were reported. In the 1950s Lunenfeld and colleagues, aware that menopausal women secreted raised levels of gonadotrophins (such as FSH), extracted gonadotrophins from menopausal urine and named it hMG (human menopausal gonadotrophin), and showed it could stimulate follicular development in female animal models. Their next step was to develop gonadotrophic preparations suitable for therapeutic use in humans. Clinical trials showed that hMG—which contained both FSH and

LH—caused expected changes in endometrial tissue and hormone excretion consistent with ovulation. In 1961 Lunenfeld reported the first successful induction of ovulation followed by pregnancies in anovulatory women using an hMG regime. These experiments provided the basis for the ovulation induction treatment still used today in non-ovulating women with irregular or absent cycles.

However, just before this time a 'chemical' alternative to these pituitary hormones had been discovered—as many drugs are—by serendipity. A drug which was known to have an anti-estrogenic effect was being tested in anovulatory women who were at a high risk of endometrial cancer. To the great surprise of the investigators, the menstrual cycles returned in nearly all these women. The study, carried out by the American endocrinologist Robert Greenblatt in 1961, showed that 28 of 36 patients with amenorrhoea found their monthly periods restored following the administration of clomiphene citrate. Today, clomiphene is still a first-line choice for the induction of ovulation in most types of patient with ovulatory problems. Although many questions still remain about how it actually works, it's a remarkable fact that more than 50 years after Greenblatt's experiments, clomiphene is still the world's most prescribed drug in fertility care.

The absence of periods or an irregular menstrual cycle remains a common cause of infertility today, affecting between 20 and 30 per cent of infertile women. However, amenorrhoea is usually a symptom, not a physiological condition in its own right—and, when diagnosed in association with infertility, can usually be treated. Amenorrhoea is commonly defined as 'primary' when no menstruation has occurred by the age of 16, and 'secondary' when none has occurred for six months in a woman who previously did have normal menstrual function. 'Oligomenorrhoea' describes irregular cycles and is defined as having a period once every four weeks to six months. Oligomenorrhoea may occasionally coincide with ovulation; conditions in which the ovaries fail to release an egg are described as anovulatory.

The World Health Organization (WHO) has classified anovulation in three diagnostic groups dependent largely on the levels of different hormones evident on testing: WHO group I is characterised by low gonadotrophin and low estrogen levels suggestive of some central origin to abnormal ovarian function. WHO group III is defined by low estrogen levels but raised levels of FSH, suggesting that the primary cause of ovarian dysfunction resides within the ovary itself. In such cases, ovarian stimulation will not be useful, because the high FSH levels (as at the menopause) are indicative of ovarian failure. WHO group II, meanwhile, is associated with normal levels of both estrogen and FSH, suggesting no more than an imbalance in the complex endocrine system which regulates normal ovarian function.

Studies suggest that the prevalence of ovarian dysfunction in the general female population is around 10 per cent (with secondary oligo-/amenorrhoea by far more common than primary). But among those investigated in fertility clinics the prevalence is much greater, reaching as high as 25–30 per cent. The list of conditions causing amenorrhoea is long—made up of anatomical, genetic and mainly hormonal disorders. Other causes may involve endocrine failure of the hypothalamus, which may be caused by stress, excessive weight loss or exercise (as often seen in female athletes and ballet dancers). In addition, high levels of prolactin, another pituitary hormone, may disrupt the menstrual cycle. But by far the greatest cause of absent or irregular periods is explained by just one condition, polycystic ovarian syndrome, or PCOS.

Even among the general population PCOS is widespread and considered by most investigators to be the most common endocrine disorder in women of reproductive age. Stud-

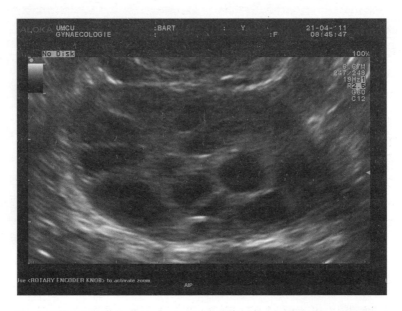

Figure 10 Multiple ovarian cysts visualised on an ultrasound scan. According to an ESHRE/ASRM consensus, polycystic ovaries are just one of three diagnostic criteria for PCOS; the others are high levels of (male) androgen hormones and irregular or absent periods. Any two of these criteria would be indicative of PCOS.

ies have found PCOS prevalence rates of between 6 and 17 per cent, depending on the diagnostic criteria used and the population studied. Moreover, although women with multiple ovarian cysts may become pregnant quite naturally, there nevertheless remain a high number who cannot, and collectively they comprise a large proportion of the patients we see in our fertility clinics; most can be safely and successfully helped to become pregnant.

Pregnancy in PCOS, however, is not simply achieved and there is growing evidence that the risk of complications to both mother and baby is clearly increased in these patients. A study we performed in Utrecht found that women with PCOS had a significantly raised risk of developing gestational diabetes and hypertension during their pregnancy, and that their babies had a higher risk of premature birth, admission to a neonatal intensive care unit and a higher perinatal mortality unrelated to multiple births. As we have seen, women with PCOS are also at a substantially increased risk of developing type 2 diabetes, hypertension and heart disease later in their lives, and thus our attention must also be given to these long-term health risks.

With some 50 per cent of PCOS cases found to be overweight, weight loss is the first advice to improve fertility, but thereafter ovulation induction with either clomiphene citrate or gonadotrophins is recommended as first- and second-line medical treatment. Fertility usually returns to normal if regular cycles can be restored. However, even though ovulation induction is a relatively simple procedure which does not require egg collection or embryo transfer, there are still distinct risks in women with PCOS—from both multiple pregnancy and OHSS. Indeed, many of the high order pregnancies reported in the press—such as the Oxford sextuplets born in 2010—are the result of ovulation induction and the over-response of the ovaries to hormonal stimulation. None of the Oxford babies weighed more than 900 grams at birth, and one—the smallest—reportedly died within 24 hours. So the risks, to both mother and babies, are high, and warrant extreme caution and constant surveillance.

However, it is also true that many of the very high order multiple pregnancies reported are not strictly 'ovulation induction', but the result of nothing more than hyperstimulation in women who are already ovulating. It is our strong view that ovulation induction treatment (with clomiphene or gonadotrophins) should be restricted to the treatment of anovulatory women; we cannot 'induce' ovulation in a woman who is already ovulating. Ovulation induction should thus be clearly differentiated from other forms of ovarian stimulation whose starting point is an ovulatory cycle and whose aim is the induction of multiple follicles (as in, for example, stimulation for intrauterine insemination).

However, the risk of multiple follicular development is strong even in strictly defined ovulation induction treatments, and in Brussels, if on ultrasound we see more than two follicles developing to maturity, we selectively aspirate the excess to reduce the risk of multiple pregnancy. This is one of several tactics in such cases: to freeze the aspirated follicles for later maturation and fertilisation; to cancel the cycle but still warn of the dangers of unprotected intercourse; or to convert ovulation

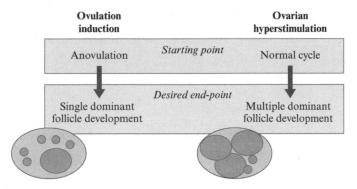

Figure 11 Ovulation induction as used in the treatment of PCOS begins with an anovulatory cycle and aims to generate a single dominant follicle (and thus a singleton pregnancy). Hyperstimulation, as in pre-treatment for intrauterine insemination, uses similar hormone therapy in a normal menstrual cycle to induce multiple follicular development with the aim of increasing the chance of pregnancy.

induction treatment to IVF, in which the fertilised eggs can be transferred singly without the risk of multiple pregnancy.

So even in women with PCOS the most common short-term health risk does indeed lie with multiple pregnancy and OHSS. This risk can be minimised by the intense monitoring of ovarian response to stimulation with ultrasound and hormonal measurement. Recent studies have shown that most anovulatory patients (around 80 per cent) given clomiphene will ovulate, providing their body weight and age are within reasonable ranges. And according to our research in Utrecht, around 50 per cent of those who do ovulate will conceive, with rates of OHSS and multiple pregnancy relatively rare in patients treated with clomiphene. However, the latest consensus advice on

PCOS recommends that clomiphene treatment should not be continued unsuccessfully beyond six cycles, after which gonadotrophins might be tried (even though clomiphene is a less expensive and more easily administered stimulation agent).

The aim of ovulation induction with gonadotrophins in women with anovulatory PCOS is to restore fertility and achieve a singleton live birth. This approach relies on the physiological concept that the growth of follicles within the ovary will be encouraged by an increase in FSH production from the pituitary prompted by the administration of gonadotrophins. However, caution is required because women with PCOS are particularly sensitive to these hormones and prone to excessive follicle development. Early experience with a high gonadotrophin starting dose (say of 150 IU FSH per day) saw unacceptably high multiple pregnancies and some cases of OHSS.

As a result, we rely today on two lower dose approaches: the first is a step-down approach developed by our group in the Netherlands in which the dosage is reduced once follicular development has been observed on ultrasound. The other, which is more commonly used, is a step-up approach which gradually increases a low initial dose until follicle growth is observed. Since then, studies have consistently shown that these low dose approaches and the constant monitoring of ovarian response by ultrasound can reduce the risk of excessive stimulation and its concomitant risks of multiple pregnancy and OHSS. Current advice—as proposed by a joint consensus statement from the ASRM and ESHRE in 2007—is to cancel the cycle if more than two follicles of 16 mm or more in diameter are seen on ultrasound.[3]

The ESHRE-ASRM consensus statement also suggests that low-dose ovulation induction protocols in anovulatory PCOS women can result in the ovulation of a single leading follicle in around 70 per cent of cases, with a very low risk of multiple pregnancy. These results compare most favourably with the 36 per cent multiple rate (and 4.6 per cent OHSS rate) formerly reported from the higher dose protocols. A prospective study from our own group in Utrecht using clomiphene citrate as first-line treatment followed by gonadotrophins as second-line achieved a cumulative singleton live birth rate of 72 per cent over a two-year treatment period in anovulatory women.

However, we must re-emphasise that many of the typical characteristics of PCOS are not always favourable to such success with ovulation induction. Multiple ovarian cysts are not the only diagnostic features, and others—seen in raised levels of 'male' androgen hormones, insulin resistance and obesity—may often dispose to infertility and a poor response to ovulation induction.

Nor should we ever forget that these non-ovarian symptoms associated with PCOS are in themselves risk factors for cardiovascular disease, and should be addressed as potentially serious for the woman herself and her baby. Thus, weight loss, the medical improvement of insulin resistance (for example, with the insulin-sensitising agent metformin), and dietary modifications are important not just for improving the prospects of fertility. In ovulation induction itself, however, metformin has been shown to offer little benefit, whether in normal or overweight subjects. And in the latter—in women with a BMI, say, over 35 kg/m²—pregnancy and live birth rates will always be significantly reduced, whatever treatment approach is used.

A third-line approach for fertility treatment in PCOS—but one showing good results—is a surgical intervention to puncture the ovarian cysts by cauterisation. This approach requires a general anaesthetic, but may be preferred in some settings, when, for example, the patient is unable to visit the clinic for regular check-ups. The reported complication rates are very favourable, and pregnancy rates seem so far to be comparable. As expected reports indicate that cauterisation is associated with lower rates of multiple pregnancy.

Intrauterine insemination

Today, 'artificial insemination' usually refers to the fertility treatment of cattle, not of humans. But when applied under its umbrella title to women it includes such reproductive procedures as intravaginal, intracervical, intratubal and intrauterine insemination (IUI), the last of which is by far the most commonly used in fertility clinics. All these insemination procedures can be performed with sperm samples from the male partner or a donor. Although controversial in some societies, and restricted by law in others, IUI with donor sperm is an increasingly common option for lesbians and single women hoping to have a baby. As opinions and legislation continue to relax, such treatments will only increase in popularity.

However, the interest in donor insemination from single women and lesbians has done little so far to stop a generally declining trend in the availability of donor sperm. The UK's national database, derived from the provision of information of all treatment cycles to the HFEA, shows that the number of women having donor insemination has declined dramatically and steadily since 1992, when nearly 9,000 women were treated. By 2007 this number had fallen to just over 2,000 women treated. Much of that decline can, of course, be explained by the sudden availability of ICSI as a more successful treatment for male infertility. However, as we have already seen, the removal of payment to donors and of anonymity have also both affected the supply of donor sperm.

Most IUI treatments are performed with the partner's own sperm, the insemination procedure merely providing a greater concentration of sperm cells in a more immediate location than natural ejaculation could ever achieve. In such cases— provided that the female has open tubes—IUI is able to combine a higher concentration of male and female gametes and represents little more than a more guaranteed alternative to natural ejaculation. However, IUI in conventional male infertility—as defined by low sperm counts and poor sperm quality—whether with partner or donor sperm has been largely superseded by ICSI. Today, therefore, unexplained infertility and the treatment of single women and lesbians are the most common indications for IUI.

Data collected by ESHRE for 2006 from most (but not all) fertility clinics in 20 European countries show that IUI with partner sperm was performed in around 130,000 cycles, and with donor sperm in around 20,000 cycles. Some 15 per cent of these treatments were in women over 40, in whom pregnancy rates were on average 7 per cent per cycle. Otherwise, pregnancy rates varied between 12 and 18 per cent, an inconsistency of outcome which is also evident in the medical literature.

Today, IUI has much evolved from its early days in the 1960s and is now reliant for its outcome on modern techniques of sperm preparation and timing to coincide with ovulation. This is often medically induced with a single shot of hCG, with the insemination performed 36 hours afterwards (as also happens in most techniques of assisted reproduction). Some centres also use ovarian stimulation prior to insemination to ensure an adequate cohort of follicles, and many studies have concluded that stimulation with clomiphene or gonadotrophins improves the outcome of IUI. However, such 'hyperstimulation' also carries the added risk of multiple pregnancy. In the ESHRE data reported above, the overall twin pregnancy rates in IUI were around 10 per cent in 2006, with rates—as expected—higher in the younger age groups.

The principal sperm preparation technique prior to insemination is washing to remove seminal fluids and grading (by centrifugation) such that the most motile sperm cells are layered above the less motile. There is no consensus on how many sperm cells should be inseminated, with studies proposing a lower limit of between 3 and 10 million in each sample. This concentrated sample, held in a suspension of culture medium, is deposited in the uterus through a small catheter, usually without ultrasound guidance. Inseminations can be done once or several times, although studies have failed to show any benefit of two attempts over one.

Studies have also consistently shown that IUI in combination with ovarian stimulation can produce better pregnancy rates than IUI alone in a natural (unstimulated) cycle. However, while the overall likelihood of conception was marginally increased in these studies, the added value of the stimulation seemed limited—particularly because multiple pregnancy occurred more frequently. In one trial the multiple pregnancy rate was as high as 29 per cent, while in another (from several centres in the US) 465 women given ovarian stimulation prior to IUI had three sets of quadruplets, four sets of triplets and 17 sets of twins; six women were hospitalised with OHSS. Even though pregnancy rates were higher in the hyperstimulation group, such excesses seem an extremely high price to pay.

As with other techniques of assisted reproduction, the success of IUI with or without ovarian stimulation is largely dependent on the age of the female partner, her duration of infertility, the condition of her Fallopian tubes, and the quality of the male partner's sperm. But it also seems clear that the prospect of pregnancy is directly related to the number of ovarian follicles developing to maturity before ovulation— and this too seems directly related to multiple conceptions and pregnancy. It was for this reason that NICE, the UK's National Institute for Health and Clinical Excellence, recommended in its benchmark fertility guidelines of 2004 that in IUI 'ovarian stimulation should not be offered, even though it is associated with higher pregnancy rates than unstimulated IUI'. In our view, this do-or-don't controversy over hyperstimulation in IUI hangs on whether a modestly increased chance of pregnancy in a stimulated cycle justifies the cost of medication, the need for ovarian monitoring and the real risk of multiple pregnancy and OHSS. So far, as we proposed in a *Lancet* report in 2006, IUI in a natural cycle carries fewer health risks than IUI after ovarian stimulation and is therefore our first-choice treatment. Certainly, IVF with single or two-embryo transfer is a much safer option because the risk of a multiple pregnancy can be controlled by the number of embryos transferred.

However, even in IUI cycles with hyperstimulation, couples cannot expect any better than modest success rates. An analysis of more than 15,000 IUI cycles in the Netherlands—some stimulated, some not—found an average ongoing pregnancy rate of just 6 per cent in each treatment cycle. This rate continued in each subsequent cycle, such that the cumulative pregnancy rate had reached 30 per cent by the seventh cycle, and 41 per cent by the ninth. If patients have the resolve and the stamina, such success rates are not unreasonable, but even the Dutch investigators proposed that nine cycles is probably enough and time to try something else, such as IVF.

Endometriosis

However, the one condition where IUI with ovarian stimulation has been favourably indicated is in women with mild endometriosis. But this too is not without controversy. The NICE guidance recommended that, 'where IUI is used to manage minimal or mild endometriosis, couples should be informed that ovarian stimulation increases pregnancy rates compared with no treatment, but that the effectiveness of unstimulated IUI is uncertain'. In fact, most studies have shown that results from IUI in mild endometriosis are less successful than in unexplained infertility, for reasons that are still not fully understood. As a result, the treatment of infertility associated with endometriosis remains one of the most challenging.

Yet endometriosis is a common condition, reported in some studies to be present in as many as 25–35 per cent of infertility patients (and up to 10 per cent of the general population). The condition is characterised by the growth of endometrial tissue as lesions outside the uterus and within the reproductive tracts and pelvic area. In recent years, some clinicians and organisations like the ASRM have developed grading systems to account for the severity of the disease and the likelihood of pregnancy in those trying to conceive naturally. But whatever the grade, it is clear that endometriosis causes pain and disruption to the daily lives of many, many women, and impedes the chance of both spontaneous and assisted conception. A variety of responses have been found and proposed to explain the latter, but it seems likely that endometriosis impedes ovarian function and ovulation. In addition, it is believed that endometriosis exerts some adverse effect on the quality of the ovulated egg. Its effect, therefore, may not be confined to the uterus and implantation.

The standard hormonal treatments for endometriosis—such as extended use of the contraceptive pill or GnRH agonists—inhibit the menstrual cycle. So these approaches may not be helpful in treating women wishing to conceive. Similarly, results from studies of IUI with ovarian stimulation have been much lower in endometriosis-associated infertility than in unexplained infertility, suggesting that the best option for infertile women with mild endometriosis is IVF. Indeed, some studies have even shown that expectant management—'wait and see'—may be preferable to IUI in the likelihood of pregnancy in women with mild endometriosis at a young age.

Reproductive surgery

In the past, especially before the widespread introduction of IVF, reproductive surgery was the cornerstone of fertility treatment and indicated in women with occluded Fallopian tubes, peritubal adhesions, congenital malformations of the uterus, and lesions from pelvic infection. But today very little remains of these approaches.

However, for women whose infertility is associated with endometriosis of the more severe grades, conservative surgery to remove or destroy the endometrial lesions is widely practised. Surgery, performed by keyhole laparoscopy, aims to improve fertility potential by restoring a more normal pelvic anatomy in women with endometrial lesions (adhesions) and endometriomas. The latter are cysts arising in or around the ovary from the growth of endometrial tissue and typically contain thick brown fluid (hence known as 'chocolate cysts'). The operative laparoscopy consists of electrocautery (via a low voltage electrical probe) or laser destruction of the endometriomas and adhesions.

A randomised trial in 1997, which compared fertility outcome in 341 women with mild endometriosis, found that those given operative laparoscopy (in which all visible endometriotic lesions were destroyed or removed) were almost twice as likely to become pregnant as those given diagnostic laparoscopy alone.[4] Fifty of the 172 women who had removal or ablation of their endometriosis became pregnant compared with 29 of the 169 women in the diagnostic-laparoscopy group. A more recent study of more than 200 Italian women having operative laparoscopy for endometriosis reported a cumulative probability of pregnancy of 30 per cent at 18 months and 50 per cent at 35 months.

The surgical removal of endometriomas from the ovary in advance of IVF is, however, much more controversial. Traditional wisdom was that removal of these chocolate cysts would inevitably improve the chance of subsequent pregnancy, either spontaneous or from IVF. There were some studies supporting this view, even though the benefit from surgery seemed limited. However, a recent randomised trial designed to investigate the effect on ICSI of removing these endometriomas found no difference in terms of fertilisation, implantation and pregnancy rates between those with prior surgery and those without. As a result, one prominent investigator was led to report that the old aphorism 'when in doubt, cut it out' is now redundant in reproduction; more appropriate is an evidence-based approach which seeks to balance the advantages and complications of cyst removal prior to IVF. Right now, that balance seems to be tilting away from surgery, with growing evidence that the procedure also damages otherwise healthy ovarian tissue and may even reduce ovarian reserve by compromising the pool of ovarian follicles.

Thus, a well-received review has urged gynaecologists 'to inform their patients that surgery has less effect on endometriosis-associated infertility than previously believed', adding that the post-operative probability of conception should be described in terms of its real, not theoretical benefit. By any interpretation of the studies, this is more than likely to be modest. It thus seems that only on the grounds of safety is the removal of endometriosis cysts justified before IVF, where leakage from the cyst may have health risks for the patient.

It is also salutary to read from the editor of one of our leading reproduction journals that, in the 'traditional' treatment of infertility, 'the era of routine laparoscopic tubal surgery has now passed'.[5] Instead, he proposed, the 'surgical' field of today is not so much the pelvis under the bright lights of the operating theatre but more likely the Petri dish in the dark-room of the IVF laboratory. The indications for diagnostic laparoscopy in the infertile patient are nearly obsolete, he wrote, as more and more fertility centres turn their backs on reproductive surgery and their greater attention to increasingly successful IVF. We too agree that assisted reproduction has superseded surgery as first-line therapy for many indications, including endometriosis-associated infertility, on the evidence-based grounds of both safety and success rates.

Not everyone is happy about the demise of surgery, especially the surgeons themselves. Even diagnostic laparoscopy is branded redundant in some centres, as clinics rely on nothing more than chlamydia testing or ultrasound for assessing pelvic pathology in many of their fertility patients. And it's fair to say that many clinicians working in fertility centres today will have only modest practical knowledge of the principles of reproductive microsurgery. So what, they say, when pregnancy rates are better with IVF than with spontaneous conception after surgery? Why correct—or even diagnose—the problem if we can bypass it? Which brings us back to the 'classical' paradigm considered earlier in this chapter and the wisdom of developing a clear diagnosis underlying the infertility.

There are, of course, some indications where surgery may be useful in the female patient, but even here that indication will depend on the age of the female and her own preference once the evidence has been presented.

In Search of the Embryo Guaranteed to Implant

Throughout the history of IVF, fertility clinics have reported their results as a percentage pregnancy rate per embryo transfer. This was convenient for them, for in their records they had immediate data on transfers and immediate data on pregnancies. Live birth was less easily defined, and data less easily available. Many pregnancies were—and still are—lost to follow-up, with deliveries performed in another hospital or even in another country. Pregnancy rates, however, while convenient for the clinic, are not so relevant for the patient. They are only of academic interest when what the couple wants is a baby, born to term and in good health.

However, pregnancy rates per embryo transfer do tell us one important thing, that many embryos transferred to the uterus do not implant to become a pregnancy. This is confirmed by the data reported to ESHRE from all IVF cycles carried out in Europe, which in 2007 culminated in a pregnancy rate per transfer of 33 per cent from IVF and ICSI. Although the rate is slowly and steadily increasing year on year (from 26 per cent in 1997) the unfortunate fact remains that at least two out of every three embryos transferred in IVF and ICSI will fail to make a pregnancy in any single treatment cycle. As we have said, disappointment is the more usual outcome to a cycle of IVF.

Why do these embryos fail to implant? It seems unlikely that the explanation lies with the IVF procedure, because we see similar—or even worse—pregnancy rates in nature, where humans have a similarly modest fertility rate of around 25 per cent. The answer, presumably, must lie with human physiology and the two functions which make pregnancy possible: the embryo's development and the receptivity of the uterus.

Today, we know that successful implantation—whether following spontaneous or assisted conception—is the result of an intimate 'cross-talk' between the developing embryo and the uterus. This cross-talk depends on a complex interaction of hormones and molecules, some of which have been explored, but the fact remains that our understanding of this complex interaction is still very limited. Indeed, our best assessment of the receptive uterus is still based on hormone measurements and the visual or histological assessment of endometrial tissue. We still have no single molecular marker of the endometrium's receptivity and the likelihood of embryo implantation.

As we mentioned in chapter 3, there have been some recent advances using techniques of 'proteomics'. These are beginning to shed more light on the physiological changes which occur

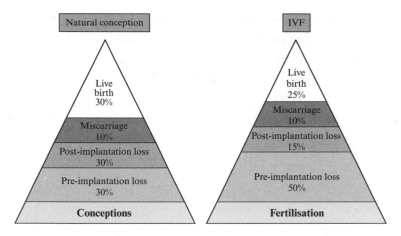

Figure 12 The delivery rate from both natural and assisted conception is roughly the same. In IVF most conceptions are lost before the embryo reaches the uterus for implantation.

in the endometrium during the menstrual cycle, and on the possible defects which may underlie infertility. Some studies have identified certain proteins which are expressed at different phases of the cycle and may yet be indicative of a receptive endometrium and the likely 'window of implantation'. These studies have been complemented by others examining the proteins present in endometrial secretions, and comparing those expressed in successful and unsuccessful cycles, thereby providing a protein fingerprint of the uterine microenvironment associated with pregnancy. The hope is that these proteomic approaches can link the molecular profiles of endometrial receptivity with successful implantation, and thus provide a uterine marker for pregnancy.

Thus, as far as implantation and the uterus is concerned, we can say little other than that the endometrium is receptive to embryo implantation during a time-restricted window when the cells of the epithelium (and its environment) are responsive. This occurs during the 'mid-secretory phase' of the cycle, which is limited to a period of around two days and is characterised by the emerging presence of several endometrial growth factors and molecules. However, we are not as yet clear what these are, and our efforts in determining the likelihood of implantation in the clinic have thus been concentrated on the embryo, which right now appears anyway to have a far greater impact on implantation success or failure.

Embryo morphology

With the universal drive to transfer fewer embryos, embryo selection has become even more important. Yet for most clinics assessment of the embryo relies on techniques which have changed very little throughout the 40-year history of IVF. For today, as ever, embryos are mainly assessed by their physical appearance as seen under the microscope in a qualitative laboratory exercise known as morphology.

It's fair to say that over the years the morphological assessment of an embryo has become a more systematic process, with clearly established grading criteria now accepted in all IVF labs. Indeed, the US Society for Assisted Reproductive Technology (SART) and European embryologists association are now trying to achieve worldwide consensus on the grading of embryos as 'good', 'fair' and 'poor'.

There are three embryonic features in a morphological assessment: the number of cells in the embryo and their rate of growth, the symmetrical shape of the embryo, and the number of cellular fragments in its composition. A good cleavage-stage (day two or three) embryo should show between four and eight cells in its development, be perfectly shaped in its symmetry, and have no more than 10 per cent of its volume filled with fragments. A poor embryo would show slow cell development, be severely asymmetric and have at least 25 per cent fragmentation.

Many studies have shown a fairly high correlation between day three embryo grading and the chance of pregnancy, with top quality embryos the most likely to implant. However, some have not, and there remains the inexplicable paradox that a large number of high grade embryos do not become viable pregnancies.

Another benefit of grading embryos at the cleavage stage is spotting the ones which look suited to extended culture in the incubator for two further days until they reach day five and the blastocyst stage. As we have seen, the transfer of a blastocyst is a more physiological process and, in most studies, associated with higher pregnancy rates. Indeed, many clinics, including ours at the VUB in Brussels, now apply a policy of single blastocyst transfer in all good prognosis patients under the age of 37 or so.

However, studies reviewing embryo morphology have suggested that morphological grading on day three is not

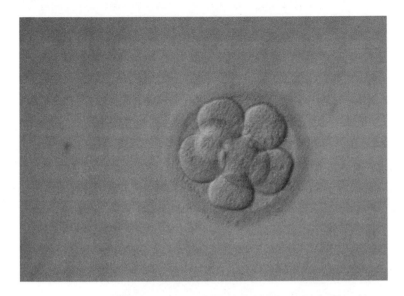

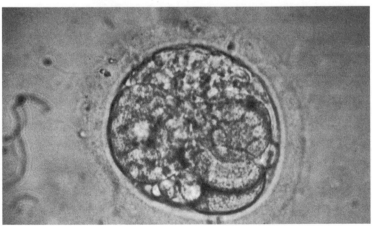

Figures 13A and 13B The top image, a good quality cleavage stage embryo with symmetrical cell development and little fragmentation. Below, this poor quality embryo shows asymmetrical cleavage, with substantial fragmentation.

necessarily associated with good blastocyst formation on day five. One important study from the Netherlands more than a decade ago showed that only 47 per cent of good quality embryos scored at day three went on to become good quality blastocysts (in contrast to 21 per cent of poor quality embryos); 45 per cent of the class 1 and 2 embryos and 69 per cent of class 3 and 4 embryos arrested in development or degenerated.[1] Similarly, a more recent study of more than 4,000 embryos morphologically assessed for cell number and fragmentation on day two found that, although transfers of a good quality blastocyst were associated with high implantation and live birth rates, the day two assessments of their early development were not helpful in predicting their implantation ability.[2]

Nevertheless, and despite progress towards a systematic method for scoring embryos, morphology remains a subjective exercise, and several studies have demonstrated high levels of variability in how individual embryologists grade their embryos. In one study, among 26 embryologists from multiple fertility centres in the US, their assessments were pitched against those of a well-regarded, internationally respected embryologist who acted as a control. Results showed that the assessments of the regular embryologists 'differed by as much as two grades' from that of the control's, despite using the same morphological grading system. The investigators concluded that such variability could affect both the number of embryos transferred and the outcome of the treatment.[3]

Our group in Utrecht also found great variability in the technique of embryo transfer and in the success of individual clinicians. We reviewed almost 4,500 transfers which inserted embryos into the uterus using two common techniques in the hands of seven physicians. The first technique—known as 'clinical touch'—had first been described by Edwards and Steptoe two years before the birth of Louise Brown in 1978 and allows the catheter to be inserted gently into the uterine cavity until it makes contact with the endometrium, whereupon it is slightly withdrawn and the embryo(s) expelled. The second method deposits the embryos at a fixed distance from the entrance to the uterus, thereby avoiding any trauma which touching the endometrium might cause. We found that with the 'clinical touch' method pregnancy rates differed greatly among the physicians, whereas after the introduction of the fixed distance technique these differences disappeared. Furthermore, the overall clinical pregnancy rate increased from 34 to 40 per cent per transfer.

Other studies have shown other variable effects from the transfer technique, which seem largely dependent on the degree of trauma which the insertion procedure might cause. Further studies have also found widely differing results according to which physician was actually doing the transfer, prompting one prominent investigator to talk about a 'physician factor' and to speculate that 'too many cooks spoil the broth'. Indeed, at the VUB in Brussels we even investigated in a large randomised trial the impact of ultrasound guidance itself on embryo transfer and found no difference in pregnancy rate between those transfers with and those without ultrasound assistance.

But as embryology stands right now—and notwithstanding the physician factor—even the most successful IVF programmes can rarely achieve an overall implantation rate above

35 per cent per embryo transferred in their routine non-selected patients. However refined the levels of laboratory quality control (air quality, culture conditions, culture media), and however high the morphological grading, two out of every three embryos do not implant.

There are, as we suggested in chapter 3, a number of possible explanations, and, as we have emphasised, an IVF pregnancy rate of 35 per cent per cycle is still comparable to nature. However, the inescapable fact is that embryo selection as determined by morphology is a crude exercise, and unfortunately the view beneath the microscope is not necessarily a true reflection of embryo competence. At some point within those few days between fertilisation and transfer a malfunction in morphological embryo selection, or in uterine receptivity, or in the transfer technique itself, or in the culture conditions, is making two-thirds of our IVF embryos non-viable.

Embryo selection represents today's major challenge in assisted reproduction, yet everyday throughout the world embryologists continue to select embryos based on little more than morphology. The latest estimates suggest that around 1 million IVF and ICSI cycles are now performed worldwide each year, and most will rely on a technique which is outdated, crude and subjective for the selection of embryos. If, however, we can identify that individual embryo which we know is viable and destined to make a pregnancy, single embryo transfer and the potential of a healthy live birth will be possible in every fertility clinic. For it is in the selection of viable, healthy embryos that the true 'designer' baby lies. This is the holy grail of embryology; is it now within our grasp?

Preimplantation genetic screening by FISH

Almost 20 years ago the embryologist Santiago Munné and colleagues in New York caused a mini-sensation by reporting that the majority of normally developing embryos obtained from older patients were chromosomally abnormal.

Munné later reported that 29 per cent of all morphologically 'normal' embryos were found to be chromosomally abnormal, and in so doing offered a glimpse of an explanation as to why so many human embryos might fail to implant. These abnormalities were evident in the numerical misarrangements of what should have been perfect chromosomal pairs. These 'aneuploid' aberrations were known to begin their occurrence around fertilisation and early embryo development, when the chromosomes fail to separate properly between the two dividing cells.

Two conclusions were clearly evident from these studies: first, that embryo morphology (as determined under the microscope) correlates only partially with the embryo's chromosomal status; and second, that the rate of aneuploidy in embryos increases with maternal age, a pattern well rehearsed in the association of age with the incidence of miscarriage or Down's syndrome. Subsequently, Munné, working with Luca Gianaroli and colleagues in Bologna, Italy, reported that around 50 per cent of embryos derived from women with a poor IVF prognosis (advanced maternal age, previous IVF failure) were chromosomally abnormal. They later found a higher rate of aneuploidy in women with a history of miscarriage than in those with healthy live births. Chromosomal integrity (or its lack) thus appeared the key to embryo competence.

The technology which made these studies possible was a chromosome visualisation technique known as fluorescence in situ hybridisation (FISH) in which genetically labelled fluorescent probes bind to those parts of the chromosome with which they have a common sequence similarity. A single cell, known as a blastomere, was removed by biopsy from a four-cell embryo and analysed by FISH to visualise its chromosomal status in their paired arrangements. Numerical aberrations in these chromosome pairs (deletions, additional copies) would be evident on the FISH image.

The first FISH reports on human embryos were based on analyses of just five chromosome pairs (XY, 13, 16, 18 and 21), all associated with failed or abnormal implantation. Even though in time the technology advanced to 12-chromosome probes, this inability to analyse all 23 chromosome pairs would in time prove the partial undoing of FISH as a means of embryo screening in IVF.

The argument of Munné and Gianaroli in these studies—and one which would form the basis of preimplantation genetic screening (PGS)—was that, if FISH could detect in embryos the chromosomal aberrations associated with poor reproductive outcome (advanced age, miscarriage, poor IVF history) and transfer only unaffected embryos, these excessive risks could be removed; thus the older woman (or the one with a record of failed IVF or miscarriage) would have the same chance of embryo implantation as a woman with a good prognosis. This concept seemed supported by a study from the same group

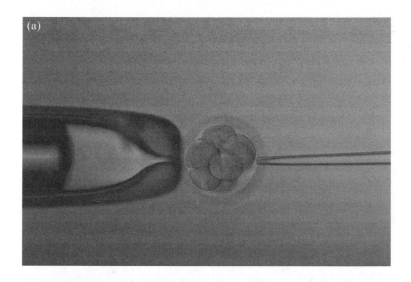

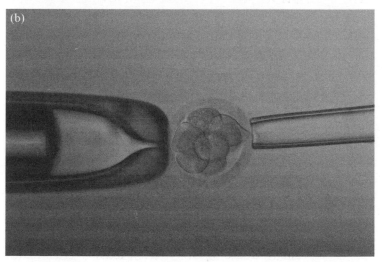

Figure 14 Continued overleaf

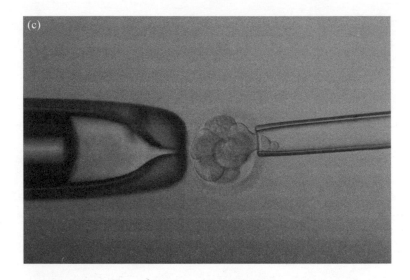

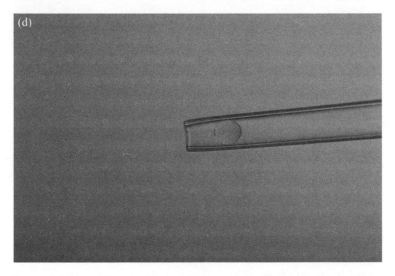

Figure 14 Biopsy from a cleavage-stage embryo. The embryo is held by a suction pipette and a single cell—known as a blastomere—is removed for analysis. Despite the apparently invasive nature of the technique, studies show that the future development of the embryo is not affected.

which found a higher rate of deliveries (and lower rate of miscarriage) in those patients whose embryos were screened by FISH than in a matched but non-PGS control group. 'From the present results,' the investigators wrote encouragingly, 'it can be postulated that aneuploidy determination prior to embryo transfer may reduce the incidence of miscarriage, and increase the rate of delivery.'

At the time, and from the results of these and similar small retrospective studies, PGS looked at best like the key to embryo selection, the opening door for single embryo transfer; at the worst it would give the older woman a better chance of pregnancy and reduce the risk of miscarriage in those susceptible. The hope was, therefore, that the exclusion of aneuploid embryos before transfer would not only improve implantation rate, but also reduce miscarriage rates and avoid the birth of children with their own chromosomal abnormalities after IVF or ICSI.

But not everyone was so hopeful or optimistic, and time would show that PGS with FISH would not improve IVF birth rates, even in older patients. Since those first PGS studies of the late 1990s, the concept has now been tested more robustly in at least 10 randomised clinical trials and not one of them has shown any benefit of PGS on pregnancy rates in IVF, even in women of an older age. Two of these trials were performed by our group in Brussels, and we found nothing to support the use of PGS in IVF.

Our first study tested the effect of PGS in two groups of women who were all at least 37 years old—thus defined as being 'of advanced maternal age'. One group had PGS, the other did not, but we found no differences in the ongoing pregnancy rates of either group. We biopsied two blastomeres from the embryos of some patients to confirm our diagnosis, and this double biopsy was later criticised as excessively invasive—and proposed as one reason for our neutral results. However, a trial we later performed to test the criticism found that either one-cell or two-cell biopsy for genetic analysis made no difference to the efficiency of FISH or to subsequent delivery rates.

Our second randomised trial of PGS was performed in women under the age of 36 who were having single blastocyst transfer. Again, we found rates of delivery to be similar in both the PGS and non-PGS groups—and we biopsied only one blastomere from the embryo. Other prospective studies from our group in Brussels have found no benefit of PGS in women with a history of miscarriage, and only the suggestion of a benefit in women with recurrent IVF failure. Other than that, PGS in our studies has proved a disappointment, with no added benefit seen to the outcome of IVF.

However, if our studies—and others—produced a neutral result for PGS, the biggest randomised trial so far, and the most controversial, actually showed that PGS had a negative effect on the outcome of IVF. The women randomised to IVF with PGS actually did worse than those randomised to IVF without PGS.[4] When the trial report was published in the *New England Journal of Medicine* in 2007, there was uproar from the supporters of PGS, especially in the US. There, without any beneficial evidence from a randomised trial, PGS had been introduced as a tried and tested means of improving embryo selection and raising the chance of pregnancy. Its practice was now widespread and attractive to patients, even at $10,000 a cycle. And suddenly, here was an upstart group from Amsterdam with only a modicum of biopsy and FISH experience saying that not only does PGS not work, but that it actually does harm to the viability of the biopsied embryo. The PGS lobby went to war, denouncing the Amsterdam study and the methods of its investigators. In different hands, they argued, PGS would work—but still there was no strong evidence to support their claim.

However, within 12 months of publication of the Amsterdam trial, both the ASRM and British Fertility Society had issued less than enthusiastic position papers about PGS. The former found no evidence to support the use of PGS for improving live-birth rates in patients with advanced maternal age, previous implantation failure and recurrent pregnancy loss, while the BFS declared that 'at present patients should be informed that there is no robust evidence that PGS for advanced maternal age improves live birth rate per cycle started'.

Nevertheless, despite such formal disapproval, PGS is still commonly offered to IVF patients, and widely performed. The latest data supplied to ESHRE show that the use of PGS may perhaps have plateaued, but there are still many cycles performed—even if some critics have argued that even to offer PGS commercially in its current form is little short of unethical. We too agree that PGS with FISH has no place in the routine use of IVF and ICSI, principally because its success depends on the erroneous assumption that the chromosomal constitution of a single blastomere is representative of the entire embryo. Our group in Utrecht has shown unequivocally that genetic mosaicism (in which each blastomere is genetically different) occurs frequently in the individual cells of early cleavage-stage embryos, thus dispelling any assumption that the chromosomal analysis of a single cell will reflect that of the entire embryo. In one study we performed in Utrecht we showed that the chromosomal status of one blastomere was indeed different from that of a second blastomere biopsied from the very same embryo. Many questions remain about the implications of these findings, particularly with respect to developmental potential. Recent observations, for example, indicate that the proportion of chromosomally normal cells increases as the embryo develops, thus suggesting some self-correcting mechanism at work. Indeed, in a further study we showed that the extent of mosaicism decreased over time with a reducing proportion of abnormal cells found in embryos of a more advanced developmental stage.

Yet the theoretical benefits of PGS remain compelling; the identification and de-selection of aneuploid embryos with little or no viability should indeed remove the risks to outcome otherwise evident in poor prognosis patients. Such approaches should allow these high-risk patients to avoid miscarriage or implantation failure and enjoy a viable normal pregnancy.

So what can be going wrong? If the theory holds good, why not the practice?

Preimplantation genetic screening by CGH

Current opinion is that there are three potential weaknesses in the PGS by FISH approach: first, that the technology cannot test all chromosomes in the biopsied blastomere; second, that the blastomere analysed may not be representative of the complete embryo (because of mosaicism); and third, that the biopsy technique itself may be harmful to the integrity of the embryo.

Now, however, some of these problems may yet be resolved by a new technique of chromosome analysis tested at two different stages of egg and embryo development. If the trials prove this new method of PGS to be advantageous, the door may open yet again towards risk-free single embryo transfer.

The analysis is performed by a technology known as comparative genomic hybridisation, or CGH, a test developed in the late 1990s as an application in cancer research to scan the entire genome for variations in DNA copy number. As used in embryology, CGH effectively compares the genomic picture of a test cell with that of a reference cell. If the reference genome is normal, then divergent increases or decreases in the test cells are visualised on microarray images to reflect the abnormality.

The application of CGH in embryology raised its head in 2007, and the following year a group from Colorado reported very high pregnancy rates following 'robust diagnosis' of blastocyst (not cleavage-stage) biopsies taken from a small number of patients who were all of advanced maternal age and thus with a poor prognosis for IVF. Application of the CGH technique allowed screening of all 23 pairs of chromosomes and produced a full screen in almost all cases. At the time of the report, 16 of 23 women screened had become pregnant, a remarkable pregnancy rate of 78 per cent. Implantation rate per embryo transferred was 62 per cent in the screened group, against a control rate of 28 per cent. A subsequent report from the same group (now in 45 patients whose average age was 37 years) found an aneuploidy rate of 51 per cent but still achieved a pregnancy rate of 82 per cent.[5]

However, these were small studies, and results have not yet been confirmed in larger controlled trials. Moreover, the test reported from Colorado needs several days to complete, so the biopsied embryos must be frozen while the analysis produces its results. However, that conundrum was resolved—as were the potentially confounding problems of mosaicism and embryo biopsy—by the application of CGH at a much earlier stage, indeed before fertilisation and before the embryo begins its cell division.

More than 20 years ago the embryologist Yury Verlinsky working in Chicago reported that analysis for identifying genetic risk need not be applied exclusively to cells biopsied from an embryo. Verlinsky had described a case in which the genetic analysis had been applied not to an embryonic cell but to a minute cell produced inside an egg at its very first stage of division before ovulation. Verlinsky showed that this tiny cell—known as the first polar body—allowed the identification of genetic mutations which, following fertilisation, would be also carried as a genetic inheritance by the embryo.

Today, 20 years later, the most exciting developments in embryo screening have now been found in CGH analysis applied to polar bodies. It's an approach which avoids all the problems inherent in embryo biopsy and FISH. First, because both polar bodies are formed in eggs (one before and one after fertilisation), a biopsied cell from an embryo (with its risk of mosaicism) is no longer necessary for the analysis. And second, because CGH provides an analysis of the full chromosomal complement, the cell is screened for abnormalities in every chromosomal pair.

The first case report describing a healthy live birth following polar body analysis of all 23 chromosome pairs by CGH appeared in early 2010. The delivery was achieved in a woman with 13 failed IVF attempts behind her and a very poor prospect of future success. Indeed, the Nottingham investigators said that women with her reproductive background had no more than a 5 per cent chance of a healthy live birth. But the use of CGH to identify normal polar bodies in an experimental series of similar patients had made possible this and four other ongoing pregnancies in what was described as 'an encouraging start' in such a poorly served group of patients.

The intriguing question, noted the investigators, 'is whether this birth heralds a novel approach that will statistically and importantly enhance the chances of live birth for all IVF patients...by reliance upon objective chromosome information instead of conventional subjective microscopic morphologic observation'. Right now, the signs are good for CGH and polar body analysis, but—as ever—its everyday worth will only be confirmed in randomised controlled trials. The proof-of-principle for such a trial has already been shown in a pilot study performed by ESHRE, which in 42 cycles tested confirmed the accuracy of the technique and its viability within a 12-hour test period (and thus without the need for freezing).

What also seems clear from this and other studies is that the biopsied cell from a cleavage-stage (day two or three) embryo may not be the most suitable target for chromosome analysis. Thus, if CGH is able to prove itself as a reliable means of genetic screening and embryo selection, it seems likely that the polar body will at least be one of the preferred cells for analysis.

The –omics

There have also been developments in identifying other cells amenable to non-invasive analysis as a marker of oocyte viability. Several studies have suggested that the layer of 'cumulus' cells (which surround the fully grown oocyte before ovulation and provide it with growth factors essential for its development) may also act as an indicator not only of oocyte competence but also of future embryo quality and pregnancy outcome. Because oocytes and cumulus cell complexes grow and develop in a co-ordinated and mutually dependent way, techniques which screen gene expression in cumulus cells may thereby provide a genetic profile associated with oocyte and embryo competence. Again, such knowledge would not only unravel some of the mystery of oocyte viability but also provide a non-invasive means of evaluating and selecting embryos.

Such techniques of analysing the activity of cells have been bundled together by reproductive scientists as '-omics'; the micro-measurement of proteins expressed by and translated from genes (especially in the endometrial environment) has become known as 'proteomics', the measurement of metabolites secreted by cells 'metabolomics', and the assessment of gene expression from oocytes and embryos 'genomics'. The idea is that the genetic activity of cells, from their DNA to the RNA produced and to the proteins and metabolites further produced, can be profiled by these -omic techniques, and in that profile a blueprint of normal cell function will emerge. Reference to this genetic blueprint can then be used to select the eggs or embryos most likely to become a successful pregnancy.

The various -omic techniques have been applied at different stages of cell development and with different targets for measurement, and they all suggest that there is indeed a 'normal' genetic micro-environment associated with the complex process of successful oocyte development, fertilisation, cell division and implantation. These are all non-invasive techniques which do not require biopsy. For example, the 'transcriptomics' applied in the screening of cumulus cells relies solely on the 'amplification' of RNA produced by thousands of active genes (to produce enough copies for detection) and their recognition (and visualisation) by sample sequences of genes on microarray chips. All the various -omic techniques seem attractive propositions as an objective alternative to morphology, but their real potential as a revolution in embryo selection will surely lie in their combined application, such that optimum viability can be assessed across the broad spectrum from ovulation to implantation.

Markers of embryo quality

It is one of the paradoxes of reproductive medicine that of two good quality embryos transferred in IVF only one implants to become a pregnancy. Why should that be? Why should one from the same group of embryos be viable and the other not? One of the most cited studies of recent years—from researchers in Melbourne and Athens—has used a combination of blastocyst biopsy and genetic analysis to identify those specific blastocysts among the same cohort which have a genetic profile consistent with viability and ongoing pregnancy. The group used a technique of DNA 'fingerprinting' to identify the embryonic origins of the babies born, and thus the composition of a genetic profile consistent with embryo viability.

The study was performed in 48 IVF patients whose oocytes were routinely fertilised and cultured to the blastocyst stage. Between eight and 20 cells from the outer (trophectoderm) cell layer of all the resulting blastocysts were then removed by biopsy. These samples were amplified and their gene expression analysed. One or more blastocysts were then transferred to all 48 women, 25 of whom became pregnant, with 37 babies born. In seven of these successful cases all the transferred blastocysts implanted, but in 18 of them some implanted and some did not.[6]

Once the babies were born, blood from the umbilical cord or swabs of cheek cells were taken and stored. The investigators then applied DNA fingerprinting to these samples to match them with the DNA obtained from the earlier trophectoderm biopsies, thereby identifying exactly which embryos grew into which babies. The genetic analysis identified which genes were expressed in the viable, successful blastocysts, and the study therefore made the link between blastocyst viability and gene profile. Already, certain genes known to be involved in cell growth and implantation have been identified, and the group has expressed a hope that this complex genetic profile can be refined to a small number of genes which are highly predictive of pregnancy. 'The ability to select the single most viable embryo from within a cohort available for transfer will revolutionise the practice of IVF,' said Gayle Jones, one of the study's investigators, 'not only by improving pregnancy rates but eliminating multiple pregnancies and the attendant complications.' As we have emphasised, one of the major stumbling blocks to worldwide acceptance of a single embryo transfer policy right now is our inadequate ability to select the viable single embryo. The opportunity to use objective, measurable criteria rather than the subjective observations of morphology should improve that predictive value and provide sufficient confidence for IVF clinics to perform single embryo transfers in all patients without a concomitant decline in pregnancy rates.

Thus, what this important study has shown is that embryo quality—as a determinant of viability and pregnancy—is not just about morphology, but also about cell biology and genetic composition. And currently, these emerging techniques of embryo assessment are encouraging attempts to identify and analyse those processes within a cell's development which are consistent with pregnancy.

What else affects the embryo?

Within the nucleus of the cell lie the 23 pairs of filament-like chromosomes which at specific locations carry the gene pairs inherited from both parents. And the genes, whose information is encoded and carried from one generation to the next in DNA, direct the synthesis of proteins which in turn drive the activity of the cell. The science-based techniques we have described above are all designed to identify these chromosomal, genetic and protein processes within the developing cell as a prelude to pregnancy.

Yet there is also emerging information to indicate that conventional genetics cannot explain every process of every cellular development, and that gene expression may not be solely caused by inherited DNA. It may be—as the study of 'epigenetics' suggests—that environmental factors (even in earlier generations) may be able to switch genes on and off to produce a phenotype which is not wholly dependent on the underlying DNA. Epigenetics thus contradicts the conventional view that gene expression happens only through the DNA code which passes from parents to children during the process of fertilisation.

In human reproduction the effect of epigenetics is evident in the expression of genes which have been marked ('imprinted') epigenetically in either the mother or father, such that only one (and not both parental copies) contributes to cell development. And, depending on whether this imprinted gene is inherited from the mother or the father, this may have disastrous consequences for the child's future development and risk of disease. Because imprinted genes are sensitive to environmental signals—and because there is no paired copy to mask their effect—there is speculation that the culture medium and culture conditions in which an embryo is left to grow may have an epigenetic effect in susceptible individuals. A recent report, albeit in mice, has shown that embryo culture (from the two-cell to the blastocyst stage) in each of five tested culture media resulted in a higher rate of epigenetic disorders (and thus embryonic health) than found in similarly cultured non-IVF embryos. As a result of the study the Canadian investigators recommended 'that time in culture and number of ART procedures should be minimized to ensure fidelity of genomic imprinting during preimplantation development'.[7] This raises the possibility that even the extended culture needed for blastocyst transfer may have a harmful epigenetic effect.

However, so far most epigenetic concerns have focused on the relationship between two serious disorders each associated with imprinted genes, Beckwith-Wiedemann and Angelman syndromes. The background—and biological plausibility—to these concerns was found in assisted reproductive experience with cattle, where oocyte injection techniques or ovarian stimulation or embryo culture on rare occasions was thought responsible for pregnancy outcome known as large (or small) calf syndrome; this was subsequently attributed to an abnormality of an imprinted growth factor gene. The reproductive scientists who were later responsible for the birth of Dolly the sheep also reported similar aberrations in their experimental embryos following the transfer of an adult cell nucleus into a denucleated oocyte.

Beckwith-Wiedermann syndrome (BWS) is a rare but serious 'overgrowth' condition in infants (affecting many organs, including the eyes and tongue), with a reported prevalence rate of one per 13,700 live births. Several studies of children with BWS have now been reported, and most of them (but not all) show a slightly higher incidence rate in children conceived by assisted reproduction than naturally. However, a recent review has concluded that the findings could be the result of chance or bias, but noted that 'current evidence has suggested an association' with assisted reproduction. Another study, which also found a very small increased risk of BWS and Angelmann syndrome in children conceived by IVF and ICSI recommended research to 'ensure that changes in ART protocols are not associated with increased frequencies of epigenetic changes and imprinting disorders in children born after ART'.

However, if the epigenetic effects of embryo culture on oocyte and embryo quality remain a subject for research, a little more is known of the effects of ovarian stimulation on the developing embryo. Several studies, including some from our own group in Utrecht, have already suggested that ovarian stimulation with gonadotrophins may be associated with a high frequency of chromosomal abnormality (including mosaicism) in the embryo. We have also found, after analysing ICSI embryos with FISH, a significantly higher proportion of aneuploid embryos following a conventional high-dose, long GnRH agonist stimulation protocol than after exposure to a mild, lower dose protocol. These findings have been confirmed in other studies in which it was shown that the extent of ovarian stimulation and the number of oocytes retrieved were directly correlated with the rates of aneuploidy found in the subsequent embryos.

Similarly, there is emerging evidence to suggest that the elevated hormone levels associated with ovarian stimulation interfere with the cross-talk between the competent embryo and uterus, and thereby impair endometrial receptivity. This is important, because—as we have already stressed—the relatively low rates of live birth in both assisted and spontaneous reproduction is much more to do with pregnancy loss than with conception failure.

For example, in Brussels we have already shown that ovarian stimulation can have an adverse effect on the receptivity of the uterus by 'advancing' the thickening (maturation) of endometrial tissue beyond the chronological window of implantation. Moreover, we found in this same study that the longer the duration of ovarian stimulation, the greater the degree of endometrial advancement assessed at the time of egg collection. Other studies have found comparable results, suggesting that, if the endometrial tissue is advanced three or more days beyond the chronological cycle day, implantation and pregnancy are simply not possible. Gonadotrophin stimulation seems to be the common denominator to these findings.

In our first studies on this theme the receptivity of endometrial tissue was assessed histologically; our more recent study has indicated that genes expressed by the endometrium were similarly able to discriminate between women with and without histologically advanced endometrial maturation beyond three days. It thus seems likely that clusters of genes are involved in the regulation of the implantation window, and that ovarian stimulation may—in some way or another—affect the expression of those genes. Ultimately, of course, identification of those genes—or their absence or mutation—may provide a marker of uterine receptivity.

If the hormones used to stimulate the ovary to produce multiple follicles have a deleterious effect on the uterus and the implantation window, it may be that freezing every embryo derived from the stimulation cycle and delaying transfer until the next cycle will avoid those risks. We have already seen that this 'freeze-all' policy can be useful—or indeed necessary—in women at risk of OHSS. Now, with embryo freezing techniques producing better and better results, it may well be time, for the sake of safety and efficacy, to consider the cumulative benefits of freezing all embryos from the initial stimulated cycle and delaying transfers until later natural cycles. A 'freeze-all' policy is one which we are assessing with great interest in Brussels.

Meanwhile, if as we believe the unequivocal aim of IVF and ICSI is the birth of a single healthy child, our immediate hopes must rest on the emerging scientific techniques of embryo selection to overcome what is still a widespread resistance to single embryo transfer. Regulatory authorities like the HFEA or specialist consensus groups have agreed that an acceptable multiple pregnancy rate following IVF and ICSI would be around 10 per cent, yet the latest world monitoring figures still show a twin pregnancy rate of around 20 per cent in Europe and Australia and 31 per cent in the US. The reasons for the resistance to SET are varied, but commonly include the risk of lower delivery rates, the ever increasing age of the IVF patient (and therefore fewer available eggs and a poorer chance of pregnancy) and the necessity of freezing embryos. But all those objections would be tempered if a reliable means of embryo selection and embryo transfer were available. As we will see in chapter 7, freezing techniques have been improved and widely introduced in the past few years, such that cryopreservation should now be an everyday part of the IVF programme. The success of such a programme would be measured not by pregnancy rate per transfer but by the delivery of a healthy singleton baby following a course of stress-free, cost-effective treatment tailored to the profile of each individual patient. But this gold standard of treatment will only be possible when embryo selection is based not just on the impressions of morphology, but also on the tests derived from new technologies.

Left to the judgement of embryologists, embryo selection in reality becomes nothing more than their choice of the 'best' embryo as determined by the morphological assessements available. This is what our embryologists do, all day and every day, deciding in effect where lies the greatest potential for life. To this extent, and as many crude accusations have long suggested, the embryologist is indeed 'playing God', which is one reason why some countries have banned even the notion of embryo selection. The controversial legislation introduced in Italy in 2004—though later challenged in the courts—insisted that only three embryos could be created from however many eggs were retrieved in IVF, and each of those three embryos had to be transferred, thereby removing any opportunity for selection. The real risk of triplets counted less than the moral risk of selecting one embryo at the cost of another.

Yet the truth is that right now we are limited in selecting the best embryo, because we don't know what reflects the greatest potential for life. Morphology, even new screening techniques like CGH or the -omics, may provide an ever more accurate guide, but it is still an inescapable fact that many embryos transferred in an IVF cycle fail to implant. Morphology is notoriously inaccurate, and the emerging methods are not yet clinically applicable. So we are not perfect in our selection of embryos. What we can do is make the best of what we have and eliminate those embryos which we believe are faulty. That's why making babies in the early twenty-first century cannot yet be an act of guaranteed positive selection, because the tools needed for that selection are not yet sufficiently developed. However, in CGH and the new -omics techniques designed to map gene and protein expression consistent with pregnancy, we may finally have the means to identify the embryo most likely to implant in the uterus—and, if that becomes possible, the key to single embryo transfer may well be in our hands.

Infertility Treatments for Fertile People

Chapter 5 considered embryo selection techniques for improving the chance of implantation in patients who are infertile. The ultimate aim is to increase the uptake of single embryo transfer and reduce the incidence of multiple pregnancies in the treatment of infertility. Yet the very technique used in preimplantation genetic screening (PGS)—embryo biopsy and genetic analysis—was originally developed not for the infertile but for individual couples who were known to be at risk of passing on a genetic disease to their children. These couples were not infertile, but the techniques of IVF and ICSI were appropriate to provide a pool of embryos from which one or two might be identified which were unaffected by the suspect gene. So it's something of an irony that this very same technique of embryo biopsy and genetic analysis has in its original indication in fertile couples created the most heated controversy in assisted reproduction.

The facts are that preimplantation genetic diagnosis (PGD) has, in the past 20 years, been responsible for the birth of many thousands of healthy babies who otherwise, if their conceptions had been left to nature, would have been at serious risk of inheriting a genetic disease from their parents. As we saw in chapter 1, PGD has removed the risk of cystic fibrosis in the offspring of many carriers of the genetic mutation responsible for this disease—and thus provides a perfect example of an infertility technique used for the benefit of fertile people. And there have been similar success stories in the removal of risk in many carriers of Tay-Sachs disease, especially among Ashkenazi Jews of eastern European family origin.

What is common to the application of all PGD techniques, whether to remove the risk of cystic fibrosis, Tay-Sachs or far rarer diseases, is that these patients—unlike those having PGS—are not infertile. They need fertility treatments only to provide a cohort of embryos which can each be screened for the gene mutation putatively destined to affect a future child. If the embryo is free of the mutation, it can safely be transferred, secure in the certainty that the child will also be unaffected.

PGD is just one of many 'infertility' treatments now being used in our fertile populations. Similarly, and increasingly so today, is the application of different fertility techniques to 'preserve' the fertility of young men and women before they undergo cancer treatment; at the time of treatment they are not infertile, but their chemo- or radiotherapy will probably cause such damage to the ovaries or testes that infertility will be an inevitable consequence of their treatment. Sperm and egg cryopreservation are the simple options, but there are now several reports of successful freeze-storage and transfer of ovarian tissue and the restoration of ovarian function. And if oocyte freezing works for cancer patients, why not for those women who have not yet found Mr Right and choose to buy a little time on their biological clock? Is there any difference between a medical indication for egg freezing and a social one?

Fertility techniques have also been introduced as appropriate for many more 'social' indications: to treat the 'infertility' of lesbians and single women (whose only cause of infertility is the lack of a male partner); or to treat couples in whom one (or both) partner(s) is HIV positive, and thereby avoid any risk of vertical transmission to the resulting baby. There have also been controversies over the use of post-mortem fertilisation, to use sperm cells cryopreserved before a partner's death. More commonly, there are many instances today where fertility has been restored in men with a vasectomy and who, in a new relationship, now wish to have children.

These are the stories which make today's headlines, not so much because they adopt groundbreaking fertility techniques, but because they adopt them for use in the fertile population. And nowhere has this been more sensationally chronicled than in the reporting of PGD. For it is PGD which conjures up the worrying spectre of 'designer' babies and the attempts of everyday couples to characterise their babies with something which nature might never have intended. It might just be for a baby girl in a family of boys, but on the incline of the slippery slope there are fears that these same PGD techniques might screen for traits and not disease. Blue eyes ... blonde hair ... and the brain of Einstein.

But the fact is that traits are difficult—even impossible—to screen for. There is no single gene, or collection of single nucleotide polymorphisms (SNPs), which will definitely encode for eye colour or intelligence. Even such a common and devastating condition as heart disease is right now proving elusive as a target for gene profiling, simply because it is so multifactorial in its phenotype and cause. Breast cancer, because its risk is more commonly associated with the presence of just one or two genes, may be a more amenable target for PGD, but, even if the BRCA1 and 2 genes can be identified in the preimplantation embryo, does the possible (not certain) risk of disease in later life justify the time and cost of PGD (as well as the risk that every embryo will be affected, so no transfer might take place).

In our view PGD is a way to remove the inherited risks of an abnormal pregnancy, not to add a new modicum of trait to an otherwise healthy pregnancy. PGD thus sets a benchmark in assisted reproduction because of its ability to provide a healthy delivery and remove (or reduce) the risk of harm from fertilisation and pregnancy.

Preimplantation genetic diagnosis

PGD is, of course, not the only way to identify and remove the inheritable risk from a pregnancy. Tay-Sachs disease, for example, a devastating recessive genetic condition whose prevalence was 10 times more common among Ashkenazi Jews, has been virtually eliminated from orthodox Jewish communities in New York as a result of an anonymous pre-marriage screening programme from which couples contemplating engagement can learn their carrier status. Tay-Sachs, like cystic fibrosis, is an autosomal recessive disorder, which means that both partners must be carriers for the mutation to express itself as disease in their children (with a one-in-four risk). Participants in the US scheme are given a personal identification number, but not the results of their simple screening blood-test. A couple considering marriage submit their numbers to the screening bank and their match is considered compatible if both partners are not carriers of the same recessive gene. The programme was set up by the Jewish community in New York to remove the risk of Tay-Sachs within the parameters of Jewish law, which forbids termination of pregnancy.

Other prenatal methods of screening for genetic and chromosomal disease—such as amniocentesis or chorionic villus sampling (CVS)—are performed during pregnancy and are usually offered to women thought to be at risk of passing on such abnormalities or of miscarriage; termination of the pregnancy is the only way to avoid these risks if a disorder is diagnosed, an option which is ethically difficult for most couples and unacceptable to some.

PGD, as its pioneers acknowledged more than 20 years ago, reaches the same genetic conclusion as prenatal methods, but avoids the need for termination of pregnancy. PGD is thus proposed as an alternative to prenatal testing and is performed not during pregnancy (after 15 weeks for amniocentesis and 10 weeks for CVS) but on a single cell biopsied from a three-day old IVF embryo before pregnancy occurs. This involves the same embryo biopsy technique as used in PGS.

In 1990, when Alan Handyside and colleagues in London reported the first pregnancies following PGD, their successful patients were two couples at risk of transmitting a chromosomal disease known to affect only male offspring (with only a single, not paired, copy of the culprit gene on the X chromosome and thus certain to be expressed). Any risk of transmitting the 'X-linked' disease would thus be removed if only female embryos—with two X chromosomes—were transferred. The investigators reported at the time that two female embryos had been transferred following the biopsy of a single cell from a cleavage stage embryo and sexing by DNA amplification of a Y chromosome-specific repeat sequence, adding that both women were confirmed as carrying normal female twins. They noted that more than 200 recessive chromosome-linked diseases were known of, thus raising the vast potential of PGD for eliminating their risk in carrier couples.[1]

Two years later the same London group reported a similarly successful unaffected pregnancy in one of three women whose embryos had been tested for the delta F508 deletion known to be implicated in cystic fibrosis.[2] This time, the test was not to 'diagnose' according to gender, but to determine the actual status of the delta F508 mutation by specific DNA analysis. Thus, in just a matter of two or three years PGD had moved forward from diagnosing sex-linked disease (by chromosome analysis) to the possibility of screening for a multitude of single gene defects in couples known to be at risk. The technology would make safe delivery possible for many thousands of couples who would otherwise have termination of pregnancy (following amniocentesis or CVS) as their only option in the case of an affected fetus.

The accuracy of the diagnosis—then as now—was crucial to the success of PGD. Misdiagnosis would at worst mean the delivery of a tragically affected baby, or at best the discarding of a healthy unaffected embryo. And the technology to guarantee the accuracy of the diagnosis was hugely challenging, necessitating the amplification of a single copy of the target gene from no more than a minute cell biopsied from a four- or eight-cell embryo. The technique used by Handyside's group and then by many others was known as polymerase chain reaction (PCR), a process in which a single copy of a piece of DNA is amplified through repeated cycles of heating and cooling such that it replicates itself in a chain reaction. With 70 per cent of cystic fibrosis cases known to be caused by a deletion at position 508 of the cystic fibrosis transmembrane regulator gene, the ability to detect this mutation in embryonic cells by PCR amplification would prove groundbreaking in reducing the future prevalence of cystic fibrosis. Today, PCR remains the cornerstone of PGD testing for single-gene defects.

By 2007, when data were collected by ESHRE's PGD Consortium from 57 centres worldwide, annual PGD activity had grown to almost 6,000 cycles, with more than 1,200 babies born. Despite the neutral and even negative clinical trials at that time, PGS (screening, not diagnosis) still represented most of that activity, with 3,753 cycles (using FISH technology for the analysis); however, there were 1,203 cycles testing for single-gene defects in at-risk couples and 729 for chromosomal abnormalities (usually 'translocations', when a piece of one chromosome attaches itself to another). The most common single-gene diseases tested for were beta-thalassaemia and cystic fibrosis.

It has been reported that almost every known monogenic disease, some very rare, can today be detected in the preimplantation embryo, and some merely associated with single genes, such as breast and bowel cancers. In fact, the very few centres worldwide equipped to undertake the genetic analysis are now able to custom design a couple-specific test for any genetic disease provided that its mutation has been described and sequenced.

Many of these single-gene diseases, such as cystic fibrosis or Tay-Sachs, are autosomal recessive genes, implying that the genetic mutation occurs on one of the 22 non-sex chromosomes (autosomes) and that both genes in the pair must be abnormal to cause disease (those with one defective gene are 'carriers'). Thus, a baby born to parents who both carry an autosomal recessive mutation, such as a delta F508 deletion, has a one-in-four chance of inheriting the mutation from both parents and developing the disease; however, the baby has a 50 per cent (one in two) chance of inheriting one abnormal gene and becoming a carrier. At best, therefore, the risk of serious disease is 25 per cent, and for every four children born to two carrier parents one will develop the disease.

However, this risk assessment according to pedigree and mutation type is not always so categorically calculated, but is especially so in PGD for breast cancer, whose first successful diagnoses and unaffected embryo transfers took place in

2008. The cause of breast cancer is complex and still not fully understood, but is certainly related to genetic mutations in a number of genes responsible for controlling cell growth and repairing damaged DNA. Among these genes, particularly evident among those with inherited forms of breast cancer, are BRCA1 and BRCA2. A woman's lifetime risk of developing breast (and/or ovarian) cancer is greatly increased (by up to 80 per cent) if she inherits (from either parent) a harmful mutation in either of these two genes. This risk is often reflected in the number of close family members who are diagnosed with breast cancer.

However, the presence of a BRCA1 genetic mutation in an embryonic cell is different from that of a delta F508 mutation because the latter will certainly cause fatal disease, while the former might not. Breast cancer is not inevitable for a woman carrying a BRCA1 mutation; cystic fibrosis is for the child derived from an embryo with two copies of the delta F508 deletion. It's the difference between probability, susceptibility and inevitability—and this is why PGD testing for breast cancer genes in embryos has become so controversial. An embryo 'diagnosed' as a carrier of the BRCA1 or 2 mutation might have developed into a perfectly healthy baby and adult. Moreover, even the transfer of an unaffected embryo does not remove the risk entirely—just being female confers a one-in-nine chance of breast cancer, and the risk is also dependent on age, environmental factors and hormones. Similarly, the disease may not be fatal, so there is the real possibility that PGD may be being performed for future 'illness' rather than to eliminate the risk of certain fatal disease. In 2003 an Australian fertility clinic announced that it had used PGD to de-select embryos carrying a gene predisposing to deafness, which prompted an outcry from some, who noted that deafness is neither life-threatening nor even a certainty for those who carry the gene.

There has been similar controversy over PGD testing for Huntington's disease, an 'autosomal dominant' condition which requires only one copy of the culprit gene for its expression. However, because the phenotype of Huntington's disease is expressed only after several decades of life (known as 'late onset'), critics have argued that PGD is ethically unacceptable because the child, albeit a carrier of the mutation, can still look forward to many years of healthy unaffected life (by which time the disease may anyway be treatable). Others argue that the mere presence of the mutation is enough to guarantee development of the disease in later life which, because of its severity, justifies testing. However, even testing brings further complications—in disclosure of the results to prospective parents, who may each be unaware of their carrier status and risk of future disease. Should they be informed of the results? A report from ESHRE has discouraged non-disclosure PGD, but genetic information about late onset diseases remains a matter of great controversy both within and outside the medical profession.

It is for these reasons that the few reported cases of PGD for the 'susceptibility' BRCA1 and 2 mutations—or the licence applications which have preceded them—have been so insistent on pedigree analysis within a family tree of close family members to signify the gravity of the risk. In the first case of PGD for BRCA1 and 2 recorded in Britain the mother of the patient reported that all her husband's female relatives had had breast cancer. Indeed, many of our own PGD patients in Brussels have seen first-hand, in close relatives and even their own babies, what devastation these genetic conditions can cause, in disability, developmental problems and premature death.

So it's important to recognise that PGD is never a simple undertaking—either for the clinic or the individual. At the VUB in Brussels, where we have a long experience of PGD (including 25 procedures for BRCA1 and 2 mutations), each case requires a close collaboration with our genetics department and extended patient counselling (both genetic and clinical). There is always the chance that every embryo tested will be affected by the target gene, and the prospect that none will be suitable for transfer. In such cases our 'fertility' treatment has actually rendered the couple 'infertile', and this is one possibility which never makes PGD a simple option. Certainly, all couples are advised that, because there are always fewer-than-usual embryos available, delivery rates are rarely as high as in conventional IVF.

In Brussels our procedure is much like that of most other centres offering PGD. The procedures of ovarian stimulation and egg collection are performed as in routine IVF patients, but in PGD cases all fertilisations are by ICSI to prevent any chance of contamination. Fertilisation is assessed 16 hours after ICSI, with further evaluation on the mornings of day two and day three. It is at this time that all embryos are morphologically graded as A, B, C or D, with the higher grades (if available) now selected for biopsy. We remove one or two blastomeres for biopsy, which shortly after we place in solution for

PCR analysis. Unaffected embryos (if any) following PCR are cultured in the incubator for a further day or two, and one or two are then transferred to the uterus as blastocysts.

However, despite the success of PGD in removing the risk of inherited disease and the birth of many healthy babies as a result, the practice remains controversial, and in some countries—notably Germany, until legislation was introduced in July 2011—it is prohibited by law. Some of these legal bans, as applied in Germany, Ireland and Italy as a function of embryo protection legislation, have recently been challenged in the courts, with equivocal outcomes, but even in countries with more liberal legislation PGD remains subject to licence and supervisory provisions. In Britain for example, the regulatory authority, the HFEA, must accept that a genetic condition is 'sufficiently serious' before it is added to the approved list of genetic diseases appropriate for PGD. In the Netherlands, PGD is only allowed in one licensed centre for an approved number of conditions. Access to PGD is thus one common reason for couples travelling overseas for 'fertility' treatment when PGD is disallowed in their own countries.

PCR remains the cornerstone of genetic analysis, but the target cell for that analysis is today less likely to be from a cleavage-stage embryo than from a blastocyst or from a polar body. Indeed, it was in the same year that Handyside described the first successful PGD that the late Yury Verlinsky and colleagues in Chicago reported that the genetic analysis of the first polar body would also allow the identification of those oocytes containing non-mutant maternal genes.

Verlinsky in 1991 described the case of a woman at high risk of transferring alpha-1-antitrypsin deficiency to her children and the identification of oocytes with no evidence of the mutant gene according to polar body analysis. Embryos from two oocytes containing the unaffected gene were transferred but no pregnancy was established. However, the accuracy of the polar body diagnosis was confirmed by PCR analysis of an oocyte which failed to fertilise. Verlinsky thus added: 'Theoretically, this technique can be applied to any genetic disorder amenable to genetic analysis using PCR.'

There have also been isolated reports in which the risk of inheriting diseases caused by mutations in the DNA of mitochondria have been removed by transplanting the genetic material of one fertilised oocyte into another without transferring the mitochondria. Mitochondria, the so-called 'power houses' of a cell, have their own DNA; mitochondrial diseases, which are usually caused by mutations in the mitochondrial DNA, include a cluster of illnesses and symptoms often associated with poor growth, learning disability and loss of organ function. The British team which reported the first successful experiment in 2010 used surplus donor oocytes from which the nucleus had been removed, and into which the nuclei of the at-risk oocytes had been transferred. However, because the nucleus of the at-risk oocytes was removed immediately after fertilisation (a stage when the sperm and oocyte provide most of the parental genes but have not yet fused) very little of the cytoplasm was transferred with the nuclei, thus leaving behind almost all the mitochondria. The technique, therefore—although in this report no more than a proof of principle exercise—may allow high-risk parents to have children free of any risk of inheriting severe mitochondrial diseases.

Saviour siblings

It is now more than 10 years since the HFEA approved in principle the technique of human leukocyte antigens (HLA) tissue matching by which parents could have a disease-free baby (with the aid of PGD) whose umbilical cord blood cells might help cure another of their children with a genetic disease. The technique, which had also been pioneered by Verlinsky in Chicago and whose tissue-matched babies had been dubbed 'saviour siblings', has proved exceptionally controversial; pro-life campaigners warned that babies would now be bred not because they were wanted but for spare parts.

HLA are tissue antigens responsible for rejection following organ or tissue transplantations. For the treatment of some genetic diseases, such as beta thalassaemia, HLA identical cord blood transplantation is said to be the best therapeutic option in children. In exceptional cases, HLA matched cord blood transplantation is also an option for acquired diseases (leukaemia, for example).

In both these cases and without the availability of a tissue-matched donor, IVF in combination with genetic analysis of the embryos can be used to select an embryo for uterine transfer with a view to the delivery of an unaffected and HLA matched sibling. Following that birth, stem cells can be collected from the baby's cord blood and used for later transplantation to the affected brother or sister. The selected embryo(s) should be HLA identical in case of acquired disease, or HLA identical and unaffected (as determined by PGD) in the case of genetic diseases.

Back in Britain in 2002 there was added controversy because some (but not all) of the first licence applications to the HFEA were to cure rare diseases whose origins were not exclusively genetic. The HFEA refused one application from a family hoping to use PGD to treat diamond blackfan anaemia, a very rare, incurable blood disorder caused by a failure of the bone marrow to make red blood cells. Some cases had been attributed to a genetic mutation but the cause was mostly unknown. Thus, PGD was being used not to identify a disease-free embryo, but an embryo whose tissue (as defined by its HLA status) was compatible with that of the affected sibling and would not be rejected by the sibling's immune system. The HFEA had said that PGD could only be authorised when its purpose was to remove the risk of serious genetic disease. Diamond blackfan anaemia, though serious, was not inherited. Two years later, however, in the face of continuing applications and even legal threats, the HFEA changed its policy, saying that in certain cases PGD could be authorised for tissue-typing alone.

Meanwhile, with permission refused in the UK at the time, Michelle and Jayson Whitaker travelled to Verlinsky's clinic in Chicago where they produced nine embryos by IVF. PGD found that three were a close HLA tissue match to Charlie, their affected son, and the best two were transferred to Michelle; one implanted and James Whitaker was born, a brother to Charlie, who was then four years old. A sample of James's umbilical cord blood was taken at birth and sent to the Chicago clinic for testing to confirm the tissue match with Charlie. The remaining cord blood was stored at a stem cell bank in Oxford until it was needed for transplantation the following year. Two further years later, in 2005, and after the transfusions of tissue-matched cord blood, haematologists at his local hospital confirmed that Charlie's bone marrow looked 'entirely normal', and he was 'effectively cured'.

The story of the Whitakers is not just a parable of hope over despair, but a reflection too of the difficulty and complexities faced by couples and their doctors in pursuing this course of treatment. It's also a story of the triumph of science over the failings of nature, and testimony to the use of PGD in the treatment of very rare and even non-genetic diseases. Yet it is this very use of PGD to provide tissue-matched embryos for third-party use (and not just disease-free embryos for healthy birth) which remains the source of controversy today. Critics argue that the motivation for pregnancy shifts from the usual family values to the more utilitarian values of 'spare-part' medicine. Is it morally justified, they ask, to use a young child (who cannot provide informed consent) as a 'donor' of a transplant? Is it acceptable even to conceive a child with transplantation mainly in mind (and not even the health of the child, as in conventional PGD)?

By the same argument and with the same technology, is it acceptable simply to produce stem cells (and not even a live birth with therapeutic cord blood) as a source of the regenerative therapy? The moral question of whether the production of embryos only for instrumental use (that is, the supply of stem cells) has dogged the progress of stem cell research for two decades, and denied its federal funding in the US.

In Brussels we began HLA tissue typing in 2001, shortly after Verlinsky and colleagues described their first case of PGD with

HLA antigen testing. We too accepted Verlinsky's premise that, if there is no HLA identical donor in the family and if there is no matching unrelated donor available in tissue banks, IVF in combination with PGD analysis of the embryos can be justly used to select an embryo for transfer to provide an unaffected HLA-matched sibling.

Since then we have treated 81 couples in Brussels in 230 cycles for HLA tissue typing. In our experience the most common indication for HLA typing alone (for acquired disease) is leukaemia, and the most common for tissue matching in combination with PGD (for inherited disease) is sickle cell anaemia. Our substantial experience shows that a conclusive HLA diagnosis can be assured in 93 per cent of the embryos tested. Our overall implantation rate is 25 per cent and the birth rate per cycle is 10 per cent. By the close of 2010, our HLA programme had resulted in the birth of 23 babies, with one pregnancy ongoing. Four successful stem cell transplantations had been completed, with some postponed and many about to begin.

We are acutely aware that the application of PGD for HLA tissue typing has raised ethical objections, particularly the selection (and possibly destruction) of embryos on the basis of a characteristic which does not actually affect the health of the baby. However, in our view HLA typing can be justified by the fact that it might well save the life of another child. The second ethical objection is that the future child is created as an instrument to cure another. However, if there already existed in the family an HLA-matched sibling, we believe it would be acceptable to consider that child as a stem cell donor. And in our experience, and in view of the extraordinary efforts made by parents to save their sick children, we believe they would never consider their HLA-matched child as anything other than the equal of their existing child.

Brussels remains just one of very few centres in the world offering HLA typing on human embryos, and we, like the other centres, have considered these ethical arguments very carefully. We continue to condemn any embryo selection on the basis of non-pathological characteristics, and our purpose is only to prevent harm to (even save the life of) any existing or future child. Belgian law on medically assisted reproduction includes a special section regulating the use of PGD for HLA typing. This law stipulates two conditions: the selection must have a therapeutic benefit for an existing child of the parents, and the fertility centre must evaluate the request to make sure that the desire to have a child is not solely motivated by the therapeutic goal.

Sex selection

The PGD Consortium's data report for 2007 noted above shows that there were only 110 PGD cycles to detect X-linked diseases that year (all using FISH), the original indication of Handyside's work almost 20 years before (then using PCR). However, these same data for 2007 also showed there were almost as many cycles performed for 'social sexing'; the reasons for the sex selection were not specified but we assume the majority were for 'family balancing', an indication not allowed in almost all countries of Europe but not disallowed in the US. Indeed, US clinics openly advertise 'gender selection technology', one of them boasting that 'by examining the genetic make-up of embryos, we can virtually guarantee your next child will be the sex of your choice'. This service, they add, is 'available to nearly all patients (not just those with genetic disorders)' and that there are 'discounted travel plans' for international patients. We are also aware of sex selection by PGD for family balancing on offer from clinics in Cyprus, providing the parents already have one child.

However, the European Convention on Human Rights and Biomedicine disapproves of sex selection for any reason other than a medical indication, stating that 'the use of techniques of medically assisted procreation shall not be allowed for the purpose of choosing a future child's sex, except where serious hereditary sex-related disease is to be avoided'.

The ethical debate behind the decision is not easy to resolve, not least because attitudes vary so greatly according to religion and culture. One argument, that sex selection would cause a population imbalance (in favour of boys), seems irrelevant given the rarity of circumstances in which couples actually seek sex selection. Indeed, a widely reported study in 2006, based on a cross-sectional survey of almost 1,200 Americans, concluded that sex selection is unlikely to be used by the US population or have any impact on gender distribution.[3]

Similarly, there is an assumption, based on outdated findings from China, that most couples would want a boy and that sex selection is thus discriminatory. A 2007 study did indeed find that interest in sex selection was largely 'cultural', but by no means confined to boys; while almost all Chinese and Asian

Indian couples in 92 cycles of insemination for sex selection chose to have boys, other ethnic groups preferentially selected females.[4] The latter study, incidentally, was in US patients who did not have PGD for their sex selection but a less invasive method of intrauterine insemination with sperm which had been separated to produce samples carrying predominantly X- or Y-bearing chromosomes. One method, in which sperm cells have to swim down through a layer of albumin, assumes that Y-bearing sperms are more motile than X-bearing and therefore swim down more efficiently; results, however, remain elusive, as does the evidence that the method really works.

However, authorities (even the ASRM in the US) continue to forecast that, with the technology now widely available and accessible (wherever in the world), demand for sex selection will only increase. Indeed, a recent report in the ASRM's journal *Fertility and Sterility* has warned that blood tests to determine fetal gender (postimplantation, not preimplantation!) are now freely available on the internet, with termination of pregnancy the likely option when gender is not as desired.[5] A *British Medical Journal* report in 1992 claimed that clandestine sex selection was responsible for a total of 50 million missing females in China, and 37 million in India. The spectre of China's missing baby girls still haunts much discussion on the subject today, even though sex selection for non-medical indication is now said to be illegal in both these countries.

In Britain the HFEA has debated social sex selection with a public consultation in 2003. In a 44-page report, the HFEA concluded that its original ban of 1993 (in the first Human Fertilisation and Embryology Act) should remain in place, noting that the benefits of sex selection were 'at best debatable and certainly not great enough to sustain a policy to which the great majority of the public are strongly opposed'. The debate was thus resolved as a result of public sentiment, and it seems the British public simply didn't like the idea. In 2007, the ban was widened to include sperm sorting techniques (even with the partner's sperm), and not just PGD.

In our view there are both ethical and practical objections to sex selection, even if the aim of 'family balancing' is in theory defensible. Practically, the many nutritional or sperm selection methods claimed as successful are simply not reliably accurate. The only reliable technique is PGD, and that will necessarily imply the discarding of good quality embryos for no other reason than that they are of the wrong gender. Such an ethically unacceptable situation could only become acceptable if sex selection by PGD were combined with an egg donation programme, such that no embryos were discarded but used by those with ovarian failure whose only hope of a baby was from a donor egg.

Fertility preservation

The concept of fertility preservation for women in advance of cancer treatment is relatively new, prompted not just by the reproductive technology to make it possible but also by the hugely improved prospects of patient survival following cancer treatment. For example, age-standardised five-year survival rates in England reached 82 per cent in women diagnosed with breast cancer between 2001 and 2006—in contrast to just 52 per cent 30 years earlier. Over a similar (though slightly shorter) period, 10-year survival rates increased from 41 per cent to 73 per cent. Rates elsewhere in the world have increased comparably. A recent study of worldwide cancer survival rates found that standardised five-year breast cancer survival rates in patients diagnosed during 1990–94 varied from over 80 per cent in North America, Sweden, Japan, Australia and Finland to less than 60 per cent in Brazil and Slovakia and below 40 per cent in Algeria. Most European countries had rates in the 70–79 per cent range.[6]

So it is only in the past few years that women surviving their cancer treatments, particularly those with childhood or early onset cancers, have also had to face the prospect of a disease-free future with the likelihood that they could not have children. In the US, of more than 1.3 million people diagnosed with cancers in 2005, 4 per cent were under the age of 35.

Both radiotherapy and chemotherapy affect reproductive function, but how and to what extent vary greatly according to the drug (type and dose), location of the radiation field, and stage of disease. In men fertility can be compromised by the disease itself (most often seen in testicular cancer or Hodgkin's lymphoma), a decline in hormone levels, or more frequently a reduction in sperm number, motility, morphology and DNA integrity. In women fertility is most frequently affected by any treatment which decreases the number of primordial follicles (thus depleting ovarian reserve) and impairs the normal functioning of the ovaries. Premature ovarian insufficiency (or complete failure) is often the result.

However, it is difficult to predict the likelihood of infertility in either men or women as a result of cancer treatment. Even far-reaching recommendations made by the American Society of Clinical Oncology in 2006 conceded that 'fertility may be transiently or permanently affected by cancer treatment or only become manifest later in women through premature ovarian failure'.[7] Regular menstruation, the report noted, does not guarantee normal fertility because of an implied effect on ovarian reserve and a higher risk of early menopause.

A detailed analysis of the effects of different anti-tumour agents on sperm production in men found that some (at relevant doses) would cause 'prolonged azoospermia', while others would cause 'temporary reductions' in sperm count or 'unlikely effects'.

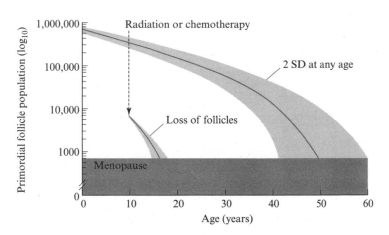

Figure 15 How cancer treatment affects the loss of ovarian reserve, in comparison with the normal loss over time. A small pool of follicles at the outset of cancer treatment may be completely depleted in a very short time.

Nevertheless, the report concluded that 'oncologists have a responsibility to inform patients about the risks that their cancer treatment will permanently impair fertility', and should refer 'interested and appropriate' patients to reproductive specialists as soon as possible. Of course, not all patients will be 'interested', and some will not be appropriate, but in our experience oncologists seem only lukewarm about referring their cancer patients to specialists in reproductive medicine. Indeed, a very recent survey, emailed to more than 1,700 oncologists in the US, found that most (95 per cent of the 249 responding) do routinely discuss the impact of cancer treatment on fertility but, although 82 per cent had referred patients to reproductive specialists, more than half did so only rarely.[8] Indeed, when planning treatment, 30 per cent said they rarely even consider a woman's concerns about fertility. Similarly, a study of cancer patients in Scotland found that very few had even been told about an effect of treatment on their fertility.

Yet the facts are, as seen in a growing number of studies, that the restoration (or maintenance) of fertility is important to cancer patients—and although most can have families through donor gametes or adoption, the majority prefer to have their own biological children. One study in men suggested that having banked sperm was a positive factor in how they coped with cancer treatment, even if the samples were never used.

The American survey cited above also showed that most of the responding physicians had only limited personal experience with fertility preservation techniques, despite their 'awareness' of them; this may partly explain their apparent lack of interest. It is clear from the reports published so far that the most advanced techniques of fertility preservation—such as oocyte or ovarian tissue cryopreservation—still seem confined to a very small group of specialist centres. However, sperm storage is a well-tested and effective means of fertility preservation for men; our more recent techniques of surgical sperm extraction from the testes or epididymis (with later fertilisation by ICSI) even make sperm banking appropriate for men with severely reduced sperm counts and motility.

As with many of these fertility preservation techniques, speed between diagnosis and treatment is a crucial factor, but even one or two semen samples collected before the start of treatment should prove enough for ICSI. Even men with testicular cancer and Hodgkin's lymphoma, whose sperm quality is often very poor, may provide enough good quality sperm cells for ICSI to be considered. Sperm banking, of course, is not possible in male cancer patients too young to provide a semen sample.

The best established option for young women is embryo freezing, but this—like oocyte freezing—requires ovarian stimulation and time, neither of which may be appropriate or possible following diagnosis. This may be especially so in very young patients (before the onset of puberty) and in those with hormone-dependent cancers; many physicians would be reluctant to expose a young woman with newly diagnosed breast cancer to the high hormone levels produced during ovarian stimulation for IVF.

Where both time and exposure to gonadotrophins present such a constraint, there is growing interest in the new techniques of in vitro maturation, where early follicles are aspirated from a natural cycle (without ovarian stimulation) and matured in culture in the laboratory before storage. So far, we have started fertility preservation in this way in some 20 patients at the VUB in Brussels, with encouraging results.

However, our most practical hopes for the preservation of female fertility now rest on oocyte cryopreservation, a procedure which has proved stubbornly difficult over the past three decades. The prospects of oocyte cryopreservation have been dramatically advanced in the past few years by the wider introduction of vitrification, a cryopreservation technique whose evidence of success is now so strong that three senior figures in the field—including Sherman Silber, a pioneer of ovarian tissue transplantation—have confidently proposed that 'this is the method of female fertility preservation that will be widely used in the near future'.[9]

However, as recently as 2006, at the annual meeting of ESHRE in Prague, Dr Masa Kuwayama from Tokyo estimated that no more than 150 babies had been born worldwide following oocyte cryopreservation by conventional 'slow-freeze' methods. But the vitrification method he pioneered and described, which requires the very rapid freezing of cells (to -196°C) in very small volumes of solution, prevents the formation of ice crystals and thereby damage to the oocyte after warming. Intra-cellular crystallisation was believed to be the main cause of damage to the oocyte's meiotic spindle and chromosome alignment once the cell was restored to normal temperature—and this was why results were so poor. Vitrification, said Kuwayama, because it preserves the oocyte in a glass-like state without crystallisation, would pave the way for a cryopreservation method as effective for female gametes as for male.

And so it has proved. In the few years since then, vitrification as a cryopreservation method for embryos, blastocysts and oocytes has been taken up with huge enthusiasm throughout the world, with research into oocyte vitrification especially driven in Italy and Germany, where embryo freezing is outlawed under current legislation. Many studies have demonstrated high fertilisation rates in warmed vitrified oocytes, with pregnancy outcomes comparable with those from fresh cycles. Indeed, a report from Spain's largest IVF group has described oocyte storage by vitrification (for oocyte banking, not fertility preservation) as more effective than traditional slow cooling, with high survival rates after warming, better preserved spindle structure and comparable DNA integrity.[10] Oocyte vitrification, they added, currently appears to be the most 'promising' method of fertility preservation in women.

Despite the encouraging results, there still remain some concerns about the safety of vitrification and the effect of the potentially cytotoxic cryoprotectants. The glass-like state which vitrification produces is made possible by extreme viscosity during the short period of temperature cooling, which in turn is produced by a high concentration (and small volume) of cryoprotectant. In the cryopreservation of reproductive cells the cryoprotectant combines high viscosity with dehydration, thereby avoiding the formation of harmful ice crystals. DMSO, a cryoportectant used at a high concentration in vitrification protocols, is known to be toxic, though so far no safety concerns have been raised in clinical trials of vitrification.

However, even oocyte cryopreservation may not be possible in some patients, such as pre-pubertal girls, for example, or women with hormone-dependent cancers. As in embryo freezing, they still need ovarian stimulation to generate an adequate supply of oocytes. In such cases the cryopreservation of ovarian tissue has been proposed and described as an alternative. This allows the long-term storage of potentially large numbers of primordial follicles which, once reimplanted after thawing, can at least theoretically restore follicular activity. The technique, however, remains experimental, and still challenging. At ESHRE's annual meeting in 2010, Claus Yding Andersen from the University Hospital of Copenhagen, currently one of Europe's most successful centres for ovarian tissue transplantation, revealed that only 14 children worldwide had thus far been born following transplantation of the frozen/thawed tissue, some of them from natural conceptions and some from IVF.

The first successful transplantation was reported in 2004 from the group of Jacques Donnez from the Catholic University hospital of Louvain, Belgium. Following the re-transplantation of thawed ovarian tissue, they described how hormone levels five months later were consistent with regular ovulatory cycles and, at 11 months, consistent with a spontaneously conceived pregnancy—which subsequently resulted in a live birth.[11]

In Denmark, where Andersen's laboratory is the only centre offering the service, a total of 16 women were reported to have had transplantations on 22 occasions, six of them twice, with three children born to two women (with a further clinical pregnancy reported). There were two unsuccessful pregnancies and five non-conceptions. However, despite the patchy results Andersen has described ovarian tissue cryopreservation as 'a valid alternative' to oocyte and embryo freezing. Its advantages, he noted, are that the cryopreservation can be performed at short notice, that the functional unit of the ovary is preserved, and that a large number of follicles are available for future fertility restoration. Most cases in Denmark have been in patients with Hodgkin's lymphoma and breast cancer.

Guidelines for ovarian tissue transplantation have been issued by several organisations in the past few years, all of them reflecting the experimental nature of the procedure and the uncertainty of outcome. Many questions still remain over the restoration of normal ovarian function after transplantation. In Utrecht our work with Donnez's group in Belgium suggests that the cyclical hormone pattern following ovarian transplantation is certainly not normal. Moreover, experience from Denmark (and Belgium) suggests that most treated women respond poorly to ovarian stimulation for IVF, with very few eggs retrieved.

However, a report describing one of the Copenhagen cases suggests that the fertility potential derived from reimplanted ovarian tissue may be substantial. This one case was a 31-year-old woman who was diagnosed with cancer in 2004 and, prior to chemotherapy treatment, had 13 strips of ovarian tissue removed and frozen to preserve her fertility.[12] The case was billed as 'the first time in the world that a woman has had two children from separate pregnancies as a result of transplanting frozen/thawed ovarian tissue'. After thawing, six thin strips of the ovarian tissue were transplanted back on to what remained of the patient's right ovary. According to the report, this ovary began to function normally again and, after mild ovarian stimulation, she became pregnant and gave birth to her first daughter in February 2007. In January 2008 she returned to Andersen's clinic for additional IVF treatment in the hope that she could conceive again. However, a pregnancy test revealed she was already pregnant naturally, and later gave birth to a healthy girl.

Commenting on the outcome, the authors of the report (who included the patient herself) proposed that the transplantation of ovarian tissue now provides 'a relatively good chance' of restoring natural fertility, and that the cryopreservation and transplantation technique 'does not in itself harm normal ovarian function'. They added that the procedure should now be encouraged in women facing 'gonadotoxic treatment'. What was striking about the case, however, was that the original transplanted strips of ovarian tissue continued to function for at least four years—and time will tell for how long after.

Indeed, Andersen said in a press statement: 'It is an amazing fact that these ovarian strips have been working for so long, and it provides information on how powerful this technique can be. [The patient] continues to have natural menstrual cycles and, at present, is using pregnancy-preventing measures to avoid becoming pregnant again.' Her ovarian function was reported to be normal, as judged by transvaginal ultrasound, with several antral follicles of different sizes visible in the ovary.

The technique described by Andersen was with strips of ovarian tissue which had been frozen before thawing and transplantation. However, in 2008 Dr Sherman Silber and colleagues working in St Louis, US, described the restoration of fertility following the transfer of ovarian tissue without freezing and in real time. This extraordinary feat was achieved in a series of monozygotic twins, one from each pair of whom was—remarkably—suffering from a premature menopause.[13] Almost all the procedures took place with fresh tissue, with the removal of one ovary from the fertile sister and its whole (one case) or sliced transplantation to the infertile 'postmenopausal' sister taking place concurrently. The fact that the twins were monozygotic minimised any risk that the graft would be rejected by the recipient sister. Silber reported that normal ovulatory menstrual cycles were restored in all recipients by 142 days. At the time of the report (2008), six had conceived naturally (one twice) and two healthy babies had been delivered.

This, of course, was not strictly fertility 'preservation', because no fertile tissue was stored for later use (in most of the patients). But Silber noted that the oldest transplant 'functioned for 36 months', somewhat less than the four years reported from Denmark, and added that the techniques described 'bode well for application to fertility preservation'. These patients, Silber continued, were healthy twins (not cancer patients), and so had no need to delay transplantation until they were assured of long-term remission or that their frozen tissue was free of malignant cells.

Since then, Silber has updated his progress and reported to the 2010 annual meeting of the ASRM that his team had removed thin ovarian strips from 140 women, around one half of them for 'social' reasons because they simply wanted to delay their first pregnancy. Thawed strips, he said, had been replanted in 11 women, who had thus far given birth to 12 babies. His estimate was that 23 babies had so far been born following fertility preservation by ovarian retransplantation, 14 from frozen grafts (as suggested by Andersen) and nine from fresh real-time grafts in his series of monozygotic twins.

At this stage, and with ovarian tissue cryopreservation such an experimental procedure, the use of such an invasive technique for what appears to be no more than a social indication cannot yet be justified. There remain many unresolved safety concerns about the storage and reimplantation of ovarian tissue, and we believe its use should at this stage be restricted to medical (and not social) indications.

It is because of these doubts that currently oocyte vitrification looks set to be the method of greatest appeal for fertility preservation, both for medical and non-medical indications. The widespread uptake of vitrification, and its acceptability as a viable means of oocyte cryopreservation, means that fertility preservation will increasingly be offered for social reasons to a young female population keen to establish careers and finances before embarking on children and family. The few reports so far on egg freezing for social indications suggest that the lack of a partner—or at least a partner who wants children—is the main reason for seeking treatment.

Not surprisingly, egg freezing for social reasons has been controversial—not least because it once again introduces an infertility treatment to the fertile population. But there have also been concerns expressed that the outcome of such practices has not been tested in clinical trials (indeed, the data on outcome even for medical indications are not robust) and that women are being offered an unquantifiable 'insurance' against future age-related infertility which in reality exploits their vulnerability for commercial ends.

It is certainly true that most of the (relatively few) women to have frozen their eggs for the future have not yet had them thawed and transferred for pregnancy—and it is true too that we cannot say with any certainty what the likelihood of pregnancy will be once they are thawed and transferred. However, we do know from studies of oocyte vitrification in Italy, for example, that warmed oocytes perform comparably with fresh oocytes and that the chance of pregnancy would appear to be the same as in routine IVF. Outcomes from oocyte donation suggest that the age of the egg is a more important prognostic factor than the age of the recipient, so freezing at an early maternal age (although counter-intuitive in women still waiting for a partner) would theo-retically improve chances.

We began our programme of social egg freezing at the VUB in Brussels in 2008 and since then have surveyed patients for their reasons and expectations. In one study of 15 applicants for social egg freezing we found their average age was just over 38, and all wanted their eggs frozen because they had not yet found the right partner with whom to have children. All had had partners in the past, but none had had children because they had not found the right man. They had all found out about the possibility of egg freezing from the internet; before that, almost 50 per cent had thought about becoming a single mother through the use of donor sperm.

Their main reasons for egg freezing were not just insurance against future infertility—they thought it would also take pressure off finding the right partner. Their downsides were the financial cost and the use of hormones for ovarian stimulation. We also found that the average age at which they thought they would use their frozen eggs was 43.4 years, an age at which, as we have seen, spontaneous conception is notoriously difficult. A later study from our Brussels group conducted among more than 1,000 Belgian women of reproductive age found that around one-third considered themselves as 'potential social oocyte freezers'. However, just over one half would not even consider the procedure.

So from our experience in Brussels, we can say that egg freezing appears an attractive option to some women who want children but have not yet met a suitable father. The technology to meet that demand now seems efficient and available in the shape of vitrification, although we cannot yet offer any kind of guaranteed outcome. Counselling, therefore, is essential on both the question of success and the procedure itself.

In 2009 the British Fertility Society (BFS) described the live birth rate associated with oocyte freezing as 'approximately 2–3%' per frozen egg. Two years earlier—well before vitrification had been so widely introduced—the ASRM concluded that live birth rate per oocyte thawed should be quoted as 2 per cent for slow freezing and 4 per cent for vitrification. Other groups have proposed that at least 20 oocytes in storage would be necessary to give a decent chance of pregnancy (implying perhaps two or three cycles of stimulation).

It was presumably because of this uncertainty over outcome—and the costs involved—that the BFS in its 2009 review gave support to egg freezing for preserving fertility in women facing cancer treatment, but 'not as a solution to counteract age-related fertility decline'. HFEA figures show that only 78 women froze their eggs for non-medical reasons in UK clinics in 2007, but still more than double the 33 who did so the year before; their average age was 37. The HFEA—which has not formally disapproved of egg freezing for social reasons—also noted (in 2010) that around 6,000 eggs have been stored in the UK (for medical and non-medical indications, with slow freezing and vitrification), from which around 150 embryos have been created. These embryos were transferred to women in around 50 cycles, which resulted in five live births.

These are not huge numbers, nor are the success rates impressive, which is why counselling is still so necessary. However, although social egg freezing is indeed new and its real value as yet unproven, it is based on a rapidly developing technology whose results are encouraging. Even at this early stage we believe egg freezing used as insurance against (age-related) infertility can improve the lifetime chance of pregnancy in women who, through individual circumstances, must defer pregnancy into their late 30s or early 40s. Data from egg donation treatments suggest that use of eggs collected and frozen from women under 35 may more than double the chance of pregnancy for a 41-year-old woman.

Medical ethicists have also noted that social egg freezing supports equal participation by women in their careers, provides more time to find and select a partner, encourages financial stability and provides a degree of protection against premature ovarian failure. Freezing eggs also avoids some of the moral and legal objections associated with freezing embryos.

Single women and lesbians

Fertility treatment for single women and lesbians is also controversial. The source of the controversy rests not on the patient but on her child and the assumed need for a father. The debate thus pitches the interests of the child against the rights of the mother, whether she is single, lesbian or defined by some minority characteristic such as advanced age or illness.

All the evidence suggests that any concerns about the child's need for a father are misplaced. Yet many health ministries recommend—or even insist—that lesbians and single women cannot be treated. One of the greatest cross-border trends we see at the VUB in Brussels is a steady stream of lesbians from France seeking treatment, simply because the fertility treatment of lesbians (and singles) is disallowed in France and allowed in Belgium.

More than a decade ago, the Dutch government criticised several licensed centres in the Netherlands because some had withheld fertility treatment from lesbians (four centres) or from single females (eight centres). Their policies, it was proposed, were not in line with Dutch equality legislation (in force since 1994), which prohibits discrimination on the grounds of religion, ideology, political persuasion, race, sex, nationality, and sexual orientation. For their part, the clinics had refused treatment on the assumption that it is preferable for children to be born into a family with both a father and mother of comparable age and without (a predisposition for) disabilities or severe diseases. In the case of lesbians, the absence of a father has been claimed to increase the risk of gender identity confusion, which may be considered developmentally unfavourable.

Yet the truth is that none of these fears and assumptions has been confirmed by well-controlled studies. Following the controversy in the Netherlands in 2000 we analysed every relevant study of parenting in lesbian and single-parent families and found strong evidence that the psychosocial development of children (whose median age was 6.1 years) and the quality of parenting were no different from those in healthy heterosexual two-parent families after infertility treatment or natural conception.

It is true, however, that many studies have shown developmental and behavioural problems in children from single-parent families: a lower level of scholastic achievement, a greater likelihood of unwanted teenage pregnancy, and more psychological problems. But these are children from families in which the mother is single because of divorce, separation or unintended pregnancy. There is no evidence that these same problems are evident in children born to single women who have had fertility treatment. As a recent study has emphasised: 'The situation of these mothers is different from that of single mothers who have separated or divorced, or who became pregnant unintentionally, in that they are generally financially secure with good social support, and the children have not been exposed to parental conflict or family disruption.'[14]

This same research group, from the Centre for Family Research in Cambridge, has been studying single and lesbian families with children conceived by fertility treatment since the mid-1990s and found in each phase of their study (in early childhood, adolescence and most recently young adulthood) that the absence of a father from birth had no negative effect on their development. Indeed, the investigators have said, the female-headed families were found to be similar to the traditional families on a range of measures of quality of parenting and young adults' psychological adjustment. Where differences were identified between family types, these pointed to more positive family relationships and greater psychological well-being among young adults raised in the female-led homes.

It is against such a reassuring background that legislation in the UK was adapted in 2008 to recognise the family rights of single women and lesbians. An original requirement for clinics to consider 'the need for a father' was changed to a consideration of 'supportive parenting'. The change in legislation, which in 2009 also allowed both same-sex partners to be named on the birth certificate and to be treated as full and equal parents from the moment of conception, is likely to increase demand from UK lesbian couples and singles for fertility treatment.

Elsewhere in the world, however, the picture of eligibility for treatment—either by law or guidelines—is far from clear. A 2010 world 'surveillance' report produced by the International

Federation of Fertility Societies noted that there has been 'liberalization in many countries involving [assisted reproductive treatment] and single women' since 1987. The report added that a requirement for marriage 'arises from beliefs associated with Islam'; however, such a view does little to explain why lesbians and singles are not accepted in France, Germany or Italy (but are in Scandinavia and Spain).

We agree that there is no reason for any systematic denial of fertility treatment to single women or lesbians—but we do insist on their additional assessment and counselling, even if that does seem discriminatory. We have to feel confident in the patient's ability to provide appropriate parenting, even though we impose no moral contraindications to treatment. Current data from the VUB in Brussels indicate that we treat some 300 single women and lesbians a year, the majority of them from France.

Our single patients, as other studies have found, tend to be in their late 30s; lesbians are somewhat younger. It thus seems that there are considerable similarities between the single women who are now beginning to consider social egg freezing and those contemplating immediate fertility treatment. Certainly, they are both without a male partner, many from relationships which have ended, and many are unhappy with their circumstances.

The common current treatment for single women and lesbians is intrauterine insemination with or without ovarian stimulation. However, because these women are not physiologically infertile, they invariably respond well to all treatments and are at risk of OHSS and multiple pregnancy when treated with

IUI and ovarian stimulation. It was for this reason that some authorities—notably the National Institute for Health and Clinical Excellence (NICE) in the UK—have discouraged ovarian stimulation with IUI and recommended that additional cycles of treatment will more than compensate. We agree with NICE that ovarian stimulation should preferably not be offered to any patients, whatever their social background. Despite its association with slightly higher delivery rates, the price is an unacceptably higher risk of multiple pregnancy (a 22 per cent twin rate was cited by NICE).

The other dilemma for single and lesbian women opting for IUI is the source of the sperm. Most choose anonymous donor sperm as supplied by a sperm bank, which is therefore compliant with requirements for serological testing and freeze-storage. However, there is a trend, particularly in countries where the law requires disclosure of the donor's name at some point, for some lesbian and single patients to choose donation from someone they know and trust. A change of law in the UK, for example, now means that anyone who has been born from donated sperm, eggs or embryos has, at the age of 18, the right to ask the HFEA for information about the donor. Similar rules apply in the Netherlands.

Sperm banks and clinics in the UK have reported that the use of a 'known donor', independently picked and introduced within a joint treatment plan, is becoming more and more popular, notably with heterosexual men providing donor sperm to their homosexual and single female friends. Another treatment trend evident in the UK is the use of IVF and egg-sharing in the treatment of lesbians.

Surrogacy

Surrogacy is a complex reproductive process which, for its success, requires the concordance of multiple expectations, interests and possibilities. Such arrangements, because of their complexity, are often aligned and confirmed by contract, although legal challenges have in the past declared such contracts invalid in some jurisdictions.

Surrogacy involves at least three individuals: a couple unable or unwilling to deliver a healthy baby, and a surrogate 'mother' who is prepared to undertake the pregnancy and delivery on their behalf. Recent headlines have exposed 'serial' surrogates in India bearing children in return for multiple payments, raising yet again the unacceptable face of commercial surrogacy, which is illegal in many European countries, including France, Italy, and the UK. It is said there are at least 50 fertility centres in India offering commercial surrogacy services. There are also said to be links between some of these centres and clinics in Europe.

Such arrangements, especially for non-medical indications, are extremely controversial, and especially so when the arrangement breaks down. There was a sensation in the US in the late 1980s when surrogate mother Mary Beth Whitehead, who had conceived a baby by intrauterine insemination on behalf of William and Elizabeth Stern, refused to hand over the baby ('Baby M') after its birth. It took a Supreme Court decision to finally award custody to the child's genetic father (and visiting rights to the gestational mother).

Elizabeth Stern was not infertile but had multiple sclerosis, which was feared might make a pregnancy too risky. In our experience there are two justified reasons to pursue surrogacy: when a pregnancy and delivery may pose life-threatening risks (as in the Sterns' case); or when the mother is without a uterus (either after surgical removal or because of genetic defects). And there are two ways in which a surrogate mother may help: first by allowing her own eggs to be fertilised (with IUI) by the male partner's sperm (also known as 'traditional' surrogacy); or receiving the transfer of embryos created by IVF from the eggs and sperm of both parents-to-be (also known as 'gestational' surrogacy). In the latter case the surrogate mother is essentially a 'host' to the couple's own biological embryo(s), and will have no genetic link to the baby. This type of host IVF surrogacy is the favoured technique of most commercial arrangements for which—according to isolated press reports from India—surrogate mothers can receive anything up to $25,000.

However, many other reports—even from the US—suggest that money is not the motivation for most surrogates; many express a degree of altruism also seen in sperm and oocyte donors, and place great importance on family life and their own children. For some couples the inability to deliver a baby safely can be just as tragic as the inability to conceive, and surrogacy provides a solution to this rare but difficult situation. Although most arrangements are made by agencies, the couple can be still be closely involved in all the progress of their baby's gestation and may even be there for the birth.

As with so many of these legally and ethically difficult procedures, surrogacy is frequently featured in cross-border arrangements. In some countries surrogacy in any form is illegal, and commercial surrogacy illegal in most. Thus, in a country like Ukraine, which appears to have liberal legislation affecting parenthood, surrogacy 'packages' are openly advertised; one Indian clinic advertised that its package (of between $22,000 and $35,000) 'covers doctor fees, legal fees, surrogate work up, antenatal care, delivery charges, surrogate compensation, egg donor, drugs and consumables, & IVF costs'. In the UK surrogacy as a medical procedure is not regulated by the HFEA, but commercial surrogacy is illegal. Thus, it may even be (though this is not legally clear) that commercial payments to overseas surrogate mothers or clinics will remove parental rights from the paying couple.

However, the potential complexities associated with surrogacy are considerable: obstetric complications in the host mother; her refusal to hand over the baby; and the legal assignation of parenthood and citizenship, particularly when different countries with different jurisdictions are involved. There have also been isolated reports based on animal models that uterine factors in the host mother may have some effect on the developing foetus, with the birth of monozygotic twins reported. Such potential problems make extensive counselling essential for any successful surrogacy arrangement. Both the host mother ('gestational carrier') and the intended parents must undergo a rigorous cycle of health screening, cytological testing and blood typing before any treatment can begin. And of course all these factors—as well as the usual prognostics of IVF and IUI—will affect outcome as well as treatment risks.

It is because of such complexities—obstetric as well as psychological—that we choose not to perform surrogacy at the

VUB in Brussels, and why it is only done in a single licensed centre in the Netherlands.

Couples who are HIV positive

Providing fertility treatment to a couple where one or both partners are HIV positive is also a specialised and complex procedure. Left to natural conception, many pregnancies will be affected because of vertical transmission to the baby (a risk rate of 25 per cent is often quoted), although treatment with antiviral drugs may reduce this risk. Nevertheless, it remains salutary to learn that nearly all HIV infections in children are the result of mother-to-baby transmission, if not during pregnancy then during breastfeeding. Many of these cases, of course, arose in women who did not know their HIV status.

However, for those who do know their status and are thinking about pregnancy, assisted reproduction is able to provide a safe conception and opportunity for a safe pregnancy. Counselling at the outset is imperative for such couples—to determine whether natural conception is safe, whether termination of a subsequently affected pregnancy is acceptable, or whether assisted reproduction is the most appropriate course.

Although the overriding concern is the safety and health of the baby, most studies have found that the desire of HIV-affected couples to have their own family is no different from that of other couples. CREAThE, a non-profit organisation developed to improve the fertility options for HIV positive couples, has described their desire for parenthood as 'psychologically and biologically sound'. In the US, CONRAD, an independent agency of the Centers for Disease Control, shares a similar aim to improve the reproductive health of HIV couples, especially in developing countries and in the prevention of HIV transmission.

The fertility technique which appears to be preferred in such cases—and by these agencies—is IVF or IUI using a sample of semen which has been 'washed'. Most studies so far have been conducted in couples in whom only the male partner is HIV positive; thus, where the risk of 'horizontal' transmission lies with the non-infected female partner and of vertical transmission with her baby. The washing technique is one widely used in IVF and employs a technique of density gradient centrifugation of the semen sample followed by sperm 'swim up' (whereby only the most motile healthy sperm cells are able to pass through a layered solution of different densities of antibiotics and supplements). The centrifugation spins all the sperm cells to the bottom of the test-tube, encouraging only the fittest to swim back up through the solution.

In 2009 one of New York's most progressive IVF centres described their 10-year experience of treating 181 couples in whom the male partner was HIV positive with washed sperm and ICSI, although the authors also noted that fertility treatments with IUI and IVF (as well as ICSI) are 'reasonable, safe and effective'. The overall clinical pregnancy rate per embryo transfer was 45 per cent, with a delivery rate of 37 per cent. The most frequent complication was multiple gestation (41 per cent).

A retrospective study performed by the CREAThE centres and published in 2007 demonstrated that IVF and ICSI following sperm washing significantly reduced the risk of HIV transmission to the uninfected female partner in a total of 580 pregnancies from 3,315 cycles. All serology tests were negative. The results, said the investigators, support the view that IVF or IUI with sperm washing could not be denied to sero-discordant couples in developed countries and, where possible, could be integrated into a global public health initiative against HIV in developing countries.

Today, more than 15,000 cycles of IUI or IVF with washed semen have been conducted in Europe, without any reported transmission of HIV to the partner woman. Thus, most organisations, including the ASRM, have now endorsed the sperm washing technique in HIV sero-discordant couples, and, despite controversy several years ago, their treatment is gaining wider acceptance—and there now may be an even greater demand for such services. Such changes in perspective have also kept in pace with the huge advances made in the treatment of HIV cases. No longer is infection deemed a death sentence, but more a chronic disease amenable to treatment.

Indeed, a 1991 a report in the *British Medical Journal* described the fertility management of HIV positive couples as 'a dilemma'. The downside of treating such patients seemed overwhelming. Yet in time, as with all such fertility treatments for non-infertile couples, their applications have gained wider acceptance and greater evidence of benefit, and the controversies have receded. HIV positive couples, lesbians, single women, even women who cannot actually deliver a baby, have all been largely integrated into the scope of modern reproductive medicine. But not by everyone, and not everywhere. Where controversy does remain is in those jurisdictions which apply restrictive legislations (as in, for example, the embryo protection laws which necessarily prohibit embryo selection of PGD) or in jurisdictions which have no guidance at all.

PGD, as we saw in chapter 2, monumentally reflects the triumph of reproductive science over the failings of nature, yet in many cultures and religions it is simply a step too far. Such views have to be respected, but it is difficult, when we know what PGD can do and what diseases it can prevent, to see how any ban on PGD is morally acceptable. In Italy in 2004, when Parliament approved legislation forcing IVF centres to fertilise no more than three oocytes and to transfer them all, a higher rate of multiple pregnancies was inevitable. How could that be construed as safe medicine?

Yet, as we shall see in the next two chapters, society must draw lines somewhere; the controversies erupt over where those lines are drawn. And, as we shall see, the limits as to how far we can go in reproductive medicine are not so much defined by our developing technologies but by the public's attitudes towards them.

Who Pays? The Social Implications

Despite Europe's ideal of harmony, there is little consistency among European governments about who should pay for fertility treatment. Even in countries like Britain, with a strong tradition of 'nationalised' health provision, 80 per cent of fertility treatments are now said to be performed in the private sector. And this despite a recommendation from the National Institute for Health and Clinical Excellence (NICE) that three IVF cycles should be funded from the public purse. ESHRE's survey of cross-border fertility treatment found 'cost' and long waiting lists at home two prominent reasons for jetting off to clinics overseas.

The supporters of publicly funded treatment (in many countries referred to as 'reimbursed' from state insurance schemes) most usually make the case that infertility is a 'disease' and should therefore be treated like any other. Behind the claim lies the definition of the World Health Organization (WHO) of reproductive health 'that people are able to have a responsible, satisfying and safe sex life and that they have the capability to reproduce and the freedom to decide if, when and how often to do so'. Anything short of that ideal thus falls short of the WHO's 'state of complete physical, mental and social well-being'. Implicit in this, adds the WHO, is the right of couples to have access to safe and effective health services.

The WHO's definition of infertility as a disease has also been seized on by many consumer groups as the justification for better services and full reimbursement of costs among their many objectives. Infertility Network UK, for example, describes itself as 'working to improve awareness and access to treatment'. Such an aim is not misplaced in the UK, where funding has not followed the NICE guidance. Indeed, patients in one region may well be eligible for state-funded IVF while in another region nearby the local health authority has no funds available: the so-called postcode lottery. In response to 'serious financial pressure' at the close of 2010, York health authority in the north of England (along with several others) announced that 'IVF procedures will be halted for the final quarter of the financial year' in a round of cost-cutting measures which appeared to include no treatment areas other than infertility.

Similarly, but on a much larger scale, Denmark, which had thus far boasted Europe's greatest access to treatment as a result of generous state funding policies, announced in May 2010 that IVF and ICSI would no longer be provided as free public health treatment. Denmark had until then offered reimbursement for up to three cycles of treatment to all types of patient. The latest move was apparently designed to cut annual government spending by 200 million Danish krone. Yet in 2007, according to ESHRE's yearly monitoring report, 4.9 per cent of all children born in Denmark were conceived by assisted reproduction. No other country in Europe, and probably not in the world, had such a high availability of treatment as Denmark, with 2,558 cycles per million inhabitants annually. Availability in countries like Germany (757 cycles per million population), France (1,091 cycles per million) and the UK (765 cycles per million) were considerably less. 'Every school class in Denmark has two IVF children on average,' said a local spokesman for ESHRE. 'These changes will have major implications and detrimental effect for childless couples, for fertility clinics and the research environment in Denmark.'

Which is just what happened in Germany in 2004. Until then government reimbursement covered 100 per cent of the cost of four treatment cycles. But that year the system was changed such that national insurance covered only 50 per cent payment for three cycles. The change prompted a dramatic decrease in the number of treatments performed in Germany, which fell from a 2003 high of 102,000 cycles to just 56,000 the following year—and is only now struggling back towards former levels.

Of course, such generous funding as formerly seen in Denmark may encourage over-treatment. Indeed, even how 'availability' falls within a paradigm of ideal IVF provision is not yet clear. We don't actually know what the ideal availability rate is, nor what effects (social or individual) are associated with over-treatment. Nor can we say that a risk (if any) derived from-overtreatment will be commensurately offset in a social model of under-treatment.

Similarly, the question of whether infertility is a disease—and therefore worthy of publicly funded treatment—is not easily answered. Certainly, if we accept the WHO claim that 'health' is not just the absence of disease but is also a psycho-social state of well-being, that reproductive health is not just about having children but having the potential to meet one's reproductive rights to safe and fulfilling parenthood, then the treatment of infertility does indeed lay claim to social support.

However, there is an increasingly strong argument today that infertility cannot be simply labelled a 'disease' as justifica-tion for funding. Most female patients we see in our practices in Utrecht and Brussels do not have infertility whose cause is some physiological problem beyond cure. Many of them are simply getting older, and their infertility is attributed only to age. Had they tried to have a baby 10 years earlier, most would have become pregnant.

Yet even this line of argument is difficult to resolve. Indeed, many, if not most, of the diseases which our state- and insur-ance-supported services treat are age-related. Heart failure, for example, which today is by far the single biggest reason for acute hospital admission, is mainly a disease of old age whose incidence is still increasing: more cases are being identified, more people are living to advancing years, and more are sur-viving a heart attack but with damage to the heart muscle. But old age does not and should not preclude treatment. Similarly, if pregnancy is safely and reasonably possible in a woman of an older maternal age, we see no reason why she should not be treated within a state system. Only when treatment may well be unsafe (for mother and baby) and/or futile (with a very poor prognosis) should it be withheld. And age—like many other components of life—should not be a contraindication to treatment. Smoking is not a reason for not treating lung cancer, nor is skiing an exclusion from the orthopaedic ward.

Treatment in developing countries

These same arguments have been played out with even more exaggeration in developing countries, where the consequences of childlessness have a far greater impact than on personal fulfilment alone. In many communities the role of children is embedded in cultural beliefs and practice, and a woman unable to have a baby may be disinherited, ostracised, forced into prostitution, or abandoned as an unwanted wife in a polygamous relationship. In such cases infertility is associated with deep personal suffering and social exclusion, and not just with the quality-of-life considerations which drive most couples in developed countries.

There has been a growing interest in the past few years to address infertility in the developing world. Yet even the most vocal supporters acknowledge that the allocation of scarce resources to infertility, when vaccination or family planning programmes appear to have far greater need, is difficult to justify. Indeed, figures from the United Nations show unequivocally that the world's highest fertility rates are in the countries of sub-Saharan Africa, with more than 5.5 lifetime births attributed to each female. This is almost three times higher than the fertility rate of 2.1 births which demographers have calculated as the rate at which a population will replace itself, and almost four times higher than the languishing rate of 1.5 births now evident in many countries of Europe. Yet in these same over-populated regions, secondary infertility has been said to affect more than one in four women of reproductive age. A WHO-endorsed report of 2004 estimated that the prevalence of secondary infertility in Tanzania was 22.7 per cent among women aged 15–49 years; in the Ivory Coast the rate was as high as 33.5 per cent. The most common causes were sexually transmitted infections and pregnancy-related sepsis.

Of course, there are IVF centres in many cities of Africa, but many of them are based on European programmes, often with personal links through their medical staff. Costs, therefore, are usually not dissimilar to those found in Europe. Egypt and South Africa also have many advanced programmes, and their experience and expertise spill over into neighbouring regions. There are reports too that the growth of IVF in China and India is keeping pace with their burgeoning economic growth. The latest (2010) issue of the world surveillance report from the International Federation of Fertility Societies lists between 102 and 300 IVF centres in China, and 500 in India, but it seems likely that the actual numbers will be many more.

Such advances, however, will have little impact on the poorer regions of Latin America, Asia and Africa. There, the investigation and treatment of infertility remain largely neglected, its victims unrecorded and frequently stigmatised. Yet, with such grave personal consequences for infertile women, there is a strong case based on natural justice that they should be treated. And the case for treatment would be even stronger if the clinical procedures of treatment—particularly IVF—were less expensive.

Today, one basic cycle of IVF in Britain costs around £4,000 including drugs; in Belgium it is similarly around €4,000 (a small proportion of which must be paid by the patient) and in the Netherlands the cost is around €2,600 (where the cost of medication represents almost half the total cost). These are prices far beyond reach in developing countries; and many of the few who could afford it would probably travel abroad for treatment anyway. However, with the possibility of lowering the cost, and simplifying the investigations and treatments, the case for incorporating fertility into family health programmes does become more persuasive.

Some leading figures in world IVF have recently proposed that low cost IVF is possible in developing countries, claiming that, with a little basic equipment in place, an IVF cycle can be performed for as little as $300. Pilot projects in Sudan, Tanzania and South Africa suggest that much of the high-technology can be removed from IVF and expensive gonadotrophins replaced by the anti-estrogen drug clomiphene or the aromatase inhibitor letrozole for ovarian stimulation. Twins are expensive, so the emphasis should be on single deliveries, with no more than one or two embryos transferred. Costly incubators may even be replaced by capsules carried at a perfect temperature in the patient's vagina, a technique already proved feasible in small studies. In some cases IVF may not even be necessary; 500

IUI cycles performed in Khartoum, Sudan, were said to have achieved a pregnancy rate of more than 10 per cent.

Despite such flickers of hope, the story of infertility in developing countries still makes grim reading. The impact of childlessness is far greater than anything we see here in our own high-tech world. Prevention of sexually transmitted diseases and unsafe backstreet abortions will help, but experience tells us that prevention campaigns rarely meet their objectives, even in developed countries. The only real solution to the widespread infertility of the developing world are those same treatments we have pioneered and perfected in the West, with the same provisions of safety but with low-cost alternatives integrated into the framework of existing clinics and hospitals. Slowly, as the general health of the developing world improves and life expectancy increases, a start has been made, with affordable treatments now available to a very small but new layer of the world's vast infertile population. And it is indeed now widely recognised that the greatest challenge facing global infertility today is the improvement of access to treatment.

Whose responsibility?

In 2008, in a report on 'the demographic future of Europe', the European Parliament called on the European Commission to 'take into consideration the sensitive issue of infertility' and, in the twenty-sixth paragraph, 'ensure the right of couples to universal access to infertility treatment'. However, despite this recognition that the treatment of infertility might make some contribution to reviving Europe's moribund fertility rates, Parliament had little to add on how this might be achieved. The European Commission has often declared that it is up to individual member states to determine how such treatments are funded and provided, and not up to some harmonising directive from the EU. Indeed, to grasp the nettle of supporting a health policy which may have demographic benefits in the distant future requires a political decision which few politicians are prepared to make. Even in response to an imminent demographic crisis, such decisions are hugely controversial—as illustrated in France in response to President Sarkozy's proposals for pension reforms.

As we saw in chapter 3, some countries—like Spain with its baby bonus—have adopted pro-fertility policies; others, like France with its 'code de famille', have a variety of tax allowances and family provisions to encourage childbirth. So the reasons why couples are not having children are not just biological (even if partly explained by a clear trend of delayed pregnancy). Child care, working hours, paid maternity (and paternity) leave, affordable housing, family tax allowances, and nursery schooling will all combine to influence the rates of conception—as will sex education and the use of contraception. So any 'policy' on treating infertility for demographic reasons should always be part of this broader policy mix.

But the treatment of infertility has also been proposed as a legitimate self-contained population policy, with its own rightful place within that policy mix. The advantages—and effects—of such a policy have been illustrated in two recent studies. One, performed by demographers at RAND Europe,

compared the potential effect on fertility rates of IVF policies in Denmark and the UK, countries with comparable
demographic and socioeconomic profiles but widely differing
policies on the funding of IVF.[1] Denmark at the time (2004)
boasted Europe's greatest availability of IVF (with 2,106 cycles
per million inhabitants and 4.2 per cent of all babies born conceived by IVF), no doubt in reflection of a generous funding
provision which covered the cost of several cycles of treatment. By contrast, in postcode-lottery Britain, the availability
of IVF was calculated at no more than 625 cycles per million
population, with just 1.6 per cent of all babies conceived by
IVF. The RAND researchers concluded that, if the UK increased
its IVF availability from 625 cycles per million population to
2,106 cycles as in Denmark, the UK's total fertility rate would
increase by 0.04—from 1.64 to 1.68. Conversely, if Denmark
were to abandon its IVF funding policy, its total fertility rate
would decline from 1.72 to 1.65. According to RAND, that 0.04
increase in the UK's fertility rate was the equivalent of around
10,000 additional children per year—and that Denmark's sub-
sidy of IVF had helped keep its fertility rate above replacement
level. RAND thus concluded—although their conclusions have
since been challenged—that assisted reproduction 'does have
potential' to contribute to total fertility and influence population structure, and that the direct costs associated with adopting fertility treatment as a population policy are comparable
with those of existing policies commonly used by governments to influence fertility.

The second study also came from health economists (and
one IVF specialist!) and proposed that an investment of £12,931

to achieve a single live birth from IVF in the UK would actually be worth 8.5 times that amount to the UK Treasury in future tax revenue.[2] Government commitment to IVF, therefore, would pay handsome rewards in the not-so-distant future. 'The analysis,' said the investigators, 'underscores that costs to the health sector are actually investments when a broader government perspective is considered over a longer period of time.'

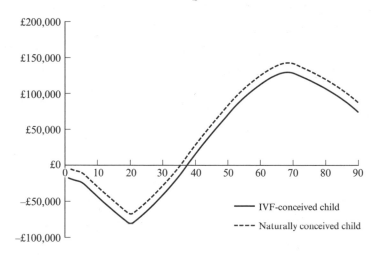

Figure 16 Health economists argued in 2009 that the projected value in terms of tax revenue to the Treasury of an IVF baby was only marginally less than a naturally conceived baby. The net balance for the state is negative while the state is providing education, health and family allowances. However, as the child enters the working years the balance shifts in favour of the state from tax payments and reduced government expenditure. The net tax position for an IVF and a naturally conceived child follow a similar pattern where the only difference between the two is the additional IVF investment cost required for conception.

These two studies presented strong arguments in favour of state-funded IVF as a population policy, but to imagine IVF reimbursement in the hands of the Treasury (and not the ministry of health) seems pure fantasy to us. Even if IVF were adopted as part of a government's responsibility for population, such a position would require a seismic shift of perspective, from the short to the long term (rarely a political option) and the relocation of infertility in the health budget priority list. Even as a quality-of-life or human rights issue, the equitable treatment of infertility as reflected in full reimbursement would still require a paradigm shift in thinking on the part of most politicians—and that we see as unlikely given the emphasis now applied to the public health priorities of heart disease, cancer, diabetes and dementia.

Indeed, no European country has as yet actually embraced IVF as a part of dedicated population policy. Even those countries which do provide IVF to their citizens as part of their nationalised health service do so as a result of a fair and equitable health policy, not a population policy. In Belgium, for example, where a scheme of six cycles of fully reimbursed IVF or ICSI treatment was introduced in 2004, that scheme was introduced as a healthcare measure dependent on the cost savings derived from single embryo transfer (obligatory in the first two cycles). The costs of the reimbursement would be more than covered by the obstetric and neonatal savings from the elimination of multiple pregnancies. This is not a population policy, but a health policy built on sound economic sense.

It thus seems unlikely that anything harder than pro-family encouragements will come out of the EU—and most likely that individual states will be left to manage their own demographic changes alone, with or without state-funded IVF. Traditionally, governments have relied on social interventions to do this (such as Spain's baby bonus), but now it does seem from the Danish model that the medical intervention of reimbursed IVF might be just as effective, or even more so. Whatever that impact is, however, it will be only small, generating no more than a modest increase in total fertility rate, but at certainly no greater cost than, say, a baby bonus.

We support state-funded IVF, but first on the grounds of health equality, not as a population policy. Certainly, we see in our clinics every week the age of our patients slowly increasing, and with it the inevitability—for some at least—of untreatable infertility. These are the women who are helping drive down fertility rates in Europe. Urging them to have their children earlier—which is central to the EU strategy—is unlikely to be effective when the economic conditions for early motherhood are not encouraging. We do believe, however, that individual governments should address the barriers to having children, especially when couples actually want children but for various reasons don't end up having them. And we continue to believe that state funding for the investigation and treatment of infertility is acceptable and desirable on health grounds, but only if the outcome is a safe pregnancy for mother and baby.

The public and the private sector

It is axiomatic that any doctor working in reproductive medicine will do his best for the patient. What that best is, however,
remains a grey area. The final chapter of this book—on the ethical debates of IVF and just how far its technology can take us—
will reiterate that the moral cornerstone of the doctor's help to
the patient is 'beneficence'—that is, to do good and, according
to the first principle of medicine, to do no harm. Yet this basic
relationship between doctor and patient in reproduction is complicated by the interests of a future child; it is the welfare of this
child which has dominated so many of the laws and regulations
introduced into reproductive medicine over the past 20 years.
Some commentators have argued that the principles of beneficence are further complicated by the paradigms of private and
public medicine: the former pays undue respect to the autonomy of the paying patient, while the latter has to accept the limi-
tation of scarce resources and the need to withhold treatment if
it seems 'futile' or merely non-cost effective to continue.

In infertility this argument is usually put in perspective by
comparing the outcomes of American and European treatment.
All IVF and ICSI in the US takes place within a private sector and
without government or state regulation (except in the conduct of
laboratory procedures). Not all patients, however, must pay from
their own pockets. Many patients, particularly those living in
states where employee health insurance is 'mandated', will have
health insurance which covers some of the investigations and

treatments. Coverage of up to three cycles of IVF is not uncommon, and the most recent estimates are that around 20 per cent of all investigation (more common) and treatment costs (less common) in infertility are covered by government mandate of private insurers or individual private health insurance.

The latest data from the US show that IVF and ICSI (using fresh non-donor embryos) achieved an overall live birth rate of 37 per cent in 2007 (and in women under 35 it was 46 per cent). This is considerably higher than the 33 per cent overall *pregnancy* rate reported by ESHRE from treatments in Europe in 2007. Implied in these results is that the American system concentrates its attention on what the patient wants, and that's the delivery of a baby from a single cycle of treatment. But the same US data also put the 'price' of that treatment into sharp perspective: 31 per cent of all IVF and ICSI births in the US in 2007 were multiple, as were 42 per cent of all births delivered to women under the age of 35 (which means that the majority of babies born from IVF in the US were delivered from multiple pregnancies). Although these rates are slowly decreasing, they still represent a far higher multiple pregnancy rate than seen in Europe, where multiple rates are now relatively steady at around 20 per cent (although this is still above the 10 per cent rate we are all aiming for).

These high multiple pregnancy rates in the US are primarily a function of the number of embryos transferred—despite the guidelines of the ASRM, which recommend a limit of one blastocyst or two embryos in patients under the age of 35 with a favourable prognosis. But, despite the guidelines—and as the graph on page 41 illustrates—in 2007 in the US only 12 per cent of all IVF/ICSI cycles involved the transfer of one embryo; 48 per cent of cycles transferred two, 33 per cent three and 14 per cent four. Admittedly, these figures do represent a declining trend in the US, but they still remain far higher than would be predicted by the ASRM's guidelines. They also ignore a warning issued by the Centers for Disease Control (CDC) in the US that assisted reproduction can present 'challenges to public health, as evidenced by the high rates of multiple delivery, preterm delivery, and low birth-weight delivery'.

The CDC also cautioned in 2010 that multiple embryo transfer in women with a good prognosis increases multiple deliveries without substantially improving live birth rates and that the trend in the US *towards* reducing the number of embryos transferred has not yet translated into a large decrease in the multiple birth rate. As in Europe, the slow reduction in the number of embryos transferred has been more than offset by improved ovarian stimulation, culture conditions and embryo selection, which have increased implantation rate and contributed to the continuing high rate of multiple births. There is also an argument that many 'difficult' patients in the older age range in the US are diverted from conventional IVF to egg donation, thereby providing a selective bias to the American IVF results.

Is it fair to say that the higher delivery and multiple pregnancy rates seen in the US are a consequence of a private treatment system? Some, including us, would say yes. In its 'world surveillance' commentary on regulations governing the number of embryos, the International Federation of Fertility Societies said: 'The need to cover costs by out of pocket expenses impacts on decisions about how many embryos to transfer...and some patients prefer twins, in spite of full knowledge of the risks of multiple pregnancy.'

Indeed, a 2008 survey of doctors practising IVF in the US found that of 52 per cent of them would deviate from the rec-ommendations of the ASRM guidelines 'for patient request'. Similar numbers said they would forego the guidelines for the transfer of frozen embryos, and 70 per cent for patients with previously failed IVF cycles—in fact, only 9 per cent said they would never deviate from the guidelines. All said they dis-cussed multiple pregnancies with their patients, but only 34 per cent reported routinely discussing single embryo transfer with all of them.[3] The survey also noted that many clinicians report 'the routine practice' of transferring two blastocysts in women under 35 and with a favourable IVF prognosis, 'a prac-tice strongly associated with twin conception after IVF, and recommended against by ASRM/SART guidelines'. If the find-ings of this single survey do indeed represent widespread prac-tice in the US, then that practice does seem—at least to us—at odds with the first principle of medicine of not doing harm.

So what the argument boils down to yet again is who decides on the number of embryos to be transferred. Those who defend the private system argue that paying patients should decide— in respect of their autonomy, a free market and privacy, but against a background of fully informed consent. But because it is society as a whole which carries the cost of multiple deliver-ies it seems reasonable to us that society should set the limits, even if enforced by law. This, it must be said, would prevent the freedom of physicians to deviate from the legislation in circumstances where it was deemed medically appropriate, but in our view 'patient request' is not an appropriate circum-stance.

Similar autonomy arguments have been raised on the subject of 'futile' treatment and whether it is ethically justified to continue the treatment of a paying patient if the chance of success is very poor. In a state-funded programme the argument is quite easily dealt with on the grounds of health economics and the cost-efficient use of limited resources. Thus, state-funded IVF in a woman of, say, 44 may be reasonably refused on the grounds that the chance of a live birth is extremely low. But what if the patient is paying out of her own pocket? What if she accepts the poor odds and still says go ahead. Or what if age is not the reason for withholding treatment, but—as we saw in chapter 3—body weight, or smoking, or excessive drinking?

In 2009 the ethics committee of the ASRM defined 'futility' in reproductive medicine as treatment with no more than a 1 per cent chance of achieving a live birth. 'Very poor prognosis' was referred to as treatment for which the odds of achieving a live birth are very low 'but not nonexistent' (that is, between 1 and 5 per cent per cycle). The ASRM ruled that withholding treatment in futile or poor prognosis cases is reasonable, but that decisions should be patient-centred (and not based on financial gain for the centre, for example). However, the ASRM did accept that 'upon request, clinicians may treat couples and/or individuals in cases of futility or very poor prognosis provided the clinician has assessed the risks and benefits and fully informed the couples and/or individuals of the low chance of success'.

In Europe the decision to treat (or not) in these difficult cases is mainly based on evidence and the statistical chance of achieving the desired result. Most clinics in Europe operating within a nationalised health service structure will impose limits on age and lifestyle. The arguments have little to do with patient autonomy or beneficence; it is simply that patients at the age of 44, or with a greatly elevated BMI, have a very slim chance of having a healthy baby by IVF.

Access to treatment

It is also a truth broadly acknowledged that to provide a fertility service only in the private sector is inherently unfair: access to treatment will depend on income, not on need. Of course, fertility is not the only area of medicine caught up in this conundrum, but it does seem to us that an infertile couple's wish to have a baby is something more than a mere lifestyle choice (as, say, cosmetic surgery or tattoo removal might be). To base access to treatment solely on ability to pay is to deny any social responsibility for treating infertility or indeed for accepting it as a condition worthy of treatment.

In 2008 ESHRE issued a position statement on 'good clinical treatment' in assisted reproduction. The idea was to provide a minimum-standards benchmark by which childless couples might get 'the best possible management of their fertility problems'. And ESHRE stated at the outset that 'accessibility' is the 'key feature' of this good clinical care. 'Treatments of proven benefit should be made easily available throughout Europe', said the report, 'irrespective of the patient's income or place of residence', adding that a lack of reimbursement policies 'constitutes a barrier to those seeking treatment'.

There seems no doubt that the last point is indisputably true. ESHRE's own IVF monitoring reports in Europe—as well as those of the CDC in the US—show that the availability of IVF (as reflected in the number of cycles performed per general population unit) is directly related to the funding background. In 2007 the two countries with Europe's highest availability of IVF (calculated as cycles per million population) were Belgium and Denmark (with 2,479 and 2,558 cycles per million respectively). These two countries were also noted for having generous policies of governmental funding for IVF. By contrast, the UK (765 cycles per million), Germany (757 cycles) and Italy (737 cycles) are known for their restrictive policies and/or declining state support.

Availability of IVF in Europe
Availability of IVF in selected European countries 2004–07, as measured by treatment cycles per million population.

	2004	2007
Denmark	2,128	2,558
Belgium	1,847	2,479
Slovenia	1,355	1,714
Sweden	1,432	1,655
France	1,154	1,091
Netherlands	942	986
UK	663	765
Germany	803	757
Italy	–	737
Portugal	–	489

And even in the free-market US the availability of IVF services is patchy—and largely dependent on reimbursement. Uptake of IVF in the US is highest in California, New York, Massachusetts, Illinois and New Jersey, but such a ranking—as in Europe—seems more determined by insurance coverage than population size alone. Using the same model as ESHRE's IVF monitoring consortium, the CDC found (for 2004) greatest availability per million population in Massachusetts (1,384), District of Columbia (1,227), New Jersey (981), Connecticut (823), and Rhode Island (790). This picture, said the CDC, was not unexpected because in 2004 Massachusetts, New Jersey and Rhode Island all had statewide mandates for insurance coverage for fertility treatment.

So the cost of IVF—as well as the regulatory environment in which it is carried out—will affect its availability. A recent analysis, rounding up costs from consumer price indices for 2006, found that the highest average price for a single cycle of fresh IVF was in the US (at €10,812), and the least expensive in Belgium and the Netherlands (at €2,450). These were the 'direct' costs of treatment, and presumably took account of consultations, drugs for ovarian stimulation, laboratory and embryology procedures, ultrasound scanning, oocyte retrieval, embryo transfer, hospital charges, nursing, counselling and overheads. Also among the cheaper nations was Japan (with an estimated charge of €3,349 per cycle), which here merits mention for an increasing use of natural cycle (or 'modified' natural cycle) IVF as a means of providing a low-cost and widely accessible service.

Japan now performs as many IVF cycles a year as the US (indeed, some assessments say more). For 2006 its IVF registry recorded almost 140,000 cycles started, which was just more than the US (124,000 cycles) and more than twice as many as the world's third IVF nation, France (with 65,000 cycles). Much of Japan's high availability of IVF is attributed to the use of milder approaches, both to enhance safety and reduce cost. This policy has been adopted with remarkable effect at the Kato Ladies Clinic in Tokyo which in 2009 alone performed almost 20,000 treatment cycles. Dr Yuji Takehara from the clinic told a medical conference in 2010 that conventional ovarian stimulation with gonadotrophins was 'quite rare' and that most cycles were 'completely natural', or stimulated only with clomiphene, or with clomiphene plus low dose gonadotrophins.

It was a serious case of ovarian hyperstimulation, said Dr Takehara, which prompted the clinic to rethink its policies. 'Since then,' he said, 'we have become suspicious about COH [controlled ovarian hyperstimulation], as fertility clinic patients are basically healthy and should under no circumstances be exposed to life-threatening side effects. That prompted us to start minimal-stimulation IVF and later revert to completely natural-cycle IVF.' The average age of the 20,000 patients was 39.4 years and pregnancy rates ranged from 13.3 per cent per transfer to 47.5 per cent depending on the protocol used. However, completely natural cycles (without ovarian stimulation) yielded high pregnancy rates, 'and we have no valid reason to rely on stimulation protocols,' said Dr Takehara. Earlier at the conference his boss, Dr Osamu Kato, had asked: Why do we retrieve so many oocytes despite the fact that only two or three are of good quality? Why do we need to transfer more than one embryo? Why are we increasing the financial burden on patients with the use of drugs?

These are questions at the heart of this book, and ones which we have tried to answer. But in the experience of this remarkable baby-making factory in Tokyo we see an answer which confirms the appeal of low-cost, minimal stimulation IVF. Cost and safety do matter, and access to treatment is restricted in a high-cost setting. The delivery rates at the Kato Ladies Clinic may not be as high as we see in the clinics of the US, but the fact is that 20,000 cycles are being performed each year in this one centre; they are reaching a cumulative pregnancy rate which is more than acceptable, at a cost which encourages repeat treatments should the first be unsuccessful, and with the minimum risk to their patients and their babies' health from the procedure itself.

This chapter is not a diatribe against the US, nor the many clinics in the rest of the world which operate in the private health sector. Our complaint is with the high cost of IVF (and unfair funding policies) which denies its widespread use in large tracts of the world's population. Ovarian stimulation has over the years become extremely complex, time-consuming, disagreeable for patients, associated with complications, and very expensive. We are now in the crazy situation in which the costs of medication for a full course of IVF (GnRH analogues, gonadotrophins, hormones for luteal support) may actually be more expensive than the cost of the IVF itself. Milder forms of ovarian stimulation will make it less complex, less time-consuming and less costly, and in so doing will improve acceptability and availability. Reported pregnancy rates per started cycle are reduced in mild stimulation protocols, but mild stimulation improves the cost effectiveness of IVF and reduces drop-out rates. It was for these very reasons that we proposed a redefinition of 'success' in IVF, away from the mantra of pregnancy and to the more relevant definition of babies born per started treatment cycle.

So we are not talking simply about 'patient-friendly' IVF, as if mild protocols and single embryo transfer are in themselves synonymous with a more sympathetic kind of treatment. Mild IVF is not the only ingredient of friendly IVF, nor is patient-friendly IVF the sole prerogative of Europe. In 2007 three specialists from New York took up the issue and argued that to define patient-friendly IVF as synonymous with minimal stimulation IVF was based on short-sighted evidence.[4] For them, genuine patient-friendly IVF lay not in the rhetoric of minimalist strategies but in the clinics which have 'responded to patient needs with sensitivity by trying to make infertility treatments more palatable'—with, for example, therapists routinely available to patients with emotional needs, computerised appointment scheduling, and satellite offices making treatment more locally accessible.

We accept and respect that view, and agree that 'mild IVF' is not an end in itself, but rather represents a complete approach to fertility which embraces both the acceptability of treatment to the patient and state, and also encourages its widespread uptake by those who need it. We also recognise—indeed, we see it over and over again in reports from US clinics—that there is a drive in the US to make IVF more socially responsible in reducing multiple pregnancy rates and making programmes less financially and emotionally demanding. The ASRM, in updating its guidelines on the number of embryos for transfer, has repeatedly emphasised the desirability of singleton deliveries. And that, we argue, is a reasonable position to take. For in our opinion, legislators, professional organisations or even individual clinics can legitimately set the same good-practice benchmarks in both the private and the public sectors, because the risks to the patient and her baby are just the same in each. Safe IVF is the objective, whoever is paying, and science should be the driver.

This was why we opposed the commercial promotion of PGS in the US, when science—the cumulative results from at least 10 different trials—said it didn't work. And even though a practice guideline from the ASRM also agreed it didn't work, PGS was still promoted as a add-on option at many US clinics. In our view, PGS with the FISH technology on offer was a futile exercise. That's why we have guidelines, to evaluate the evidence, draw a conclusion and set a benchmark for treatment whose first consideration is safety. So we believe that the state has a legitimate role to play in the safe application of our IVF technologies and under what conditions they are most reasonably used. The limits to treatment should be based on hard scientific facts, and those facts remain the same whether the patient is paying or not.

How Far Can We Go?

In October 2010 it was announced that Robert Edwards, the 'father of IVF', would be awarded that year's Nobel prize for medicine. Edwards, a reproductive biologist working alongside the gynaecologist Patrick Steptoe, had of course been responsible for the birth of the world's first IVF baby back in July 1978. Since then, the Nobel Assembly announced, 'a new field of medicine has emerged, with Robert Edwards leading the process all the way from the fundamental discoveries to the current, successful IVF therapy. His contributions represent a milestone in the development of modern medicine.' Edwards's Nobel prize was officially for 'the development of in vitro fertilization'.

That 'development' had in fact begun in the late 1950s when Edwards, working at the Institute of Animal Genetics in

Edinburgh, had successfully induced follicle growth, ovulation and pregnancy in laboratory mice in order to clarify the timing of these stages of development. Just a year or two later (now working at the National Institute for Medical Research in London) he had identified chromosome abnormalities in mouse embryos by analysing 'outgrowths of cells from pieces of tissue cultured in vitro'. Remarkably, this was achieved only a year or two after the entire chromosome numbering system had been agreed upon, and the chromosomal basis for Down's and Klinefelter syndromes established. Even more remarkable at the time, in a report to the journal Nature in 1967, Edwards and graduate student Richard Gardner had described a technique of identifying gender in five-day-old rabbit embryos, which, looking ahead to the potential of sex selection, they proposed 'could be of great importance in agriculture and human affairs'.

It was these experiments—even then—which not only encouraged Edwards to predict PGD and stem cell therapies, but also fired his own anxieties about the ethics of research on human, not mice or rabbit, embryos. Indeed, in the report to Nature in 1961 on the identification of chromosomes in his experimental mouse embryos, Edwards had been careful to note that the identification 'involved the death of the animals'. He recognised only too well that work like this, even in animal models, raised ethical concerns, and that these same concerns were sure to be exacerbated when he turned to oocyte maturation and fertilisation in humans.

The first bold headlines of outrage had appeared in 1969, when Edwards was established as Professor of Physiology of

Reproduction in Cambridge and following publication of a report from his laboratory in *Nature* on the creation of seven human zygotes from oocytes recovered laparoscopically (by Steptoe) and inseminated with sperm capacitated to achieve fertilisation in vitro. Even then the newspapers raised fears of 'test-tube babies', which prompted a statement from Edwards (as reported by the news agency Reuters) that 'we are learning to help infertile women. We do not want monstrosities.' Other London newspapers had alluded to a 'human time bomb', 'life outside the body', and 'the test-tube baby factory'. But despite the recurring Frankenstein spectres drifting through these headlines, Edwards and Steptoe never shirked from explaining and defending their work, nor from debating the ethical questions it raised.

Two years later, in a further report to *Nature*, the zygotes of the sensational 1969 paper had become eight-cell embryos; but now, rather than matured in the laboratory to the metaphase II stage of meiosis, the oocytes were aspirated by laparoscopy from the ovary immediately prior to ovulation in women stimulated with gonadotrophins. But unlike the in vitro-matured oocytes, these pre-ovulatory oocytes, when fertilised with activated sperm, cleaved beyond the zygote stage to undergo cell division consistent with blastocyst development. Now indeed, the age of the test-tube baby was about to dawn—and with it Edwards and Steptoe rose into the spotlight of the public gaze.

We now know, from the detailed research of Robert Edwards's former student Martin Johnson, that around this time Edwards and Steptoe applied to the UK's Medical Research

Council for the long-term funding of their work.[1] Steptoe, they proposed, would move from Oldham to join Edwards in Cambridge to work on a programme which combined basic and clinical research in reproductive physiology whose ultimate objective was the treatment of infertility. But the grant application was refused, as Johnson's investigation makes clear, because the MRC 'had serious doubts about the ethical aspects of the proposed investigations, especially those relating to the implantation in women of oocytes fertilized in vitro, which was considered premature in view of the lack of preliminary studies on primates and the present deficiency of detailed knowledge of the possible hazards involved'. Johnson makes clear there were other practical reasons too, but the official papers recommend that the application for support 'should be declined on ethical grounds'.

For Edwards and Steptoe it was a bitter blow, and it meant that throughout most of the 1970s—indeed until the birth of Louise Brown in 1978—they continued from their two separate bases in Cambridge and Oldham. This meant long journeys by car on a Friday night from the one to the other, no sustained funding in place, and little apparent support from the medical establishment. Not until the birth of Louise and the huge public interest she created did the MRC change its position and finally adopt a policy which accepted IVF as 'an experimental treatment' (and not 'research') and that 'human IVF with subsequent embryo transfer should now be regarded as a therapeutic procedure covered by normal doctor/patient ethics'.

The MRC, in rejecting Edwards and Steptoe's grant application, had also criticised the prominent public role which the pair had assumed as their research into oocyte fertilisation and embryo development had progressed. Indeed, Johnson provides evidence that reports of their work in *The Times* newspaper were even prepared in collaboration with the editor of *Nature*, the journal in which the research was originally published. Yet it's also clear that these forays into the popular press and television were not merely for publicity; Edwards was convinced that public understanding was a prerequisite for the acceptance of IVF and embryo transfer as a legitimate medical procedure—and that meant public education and public debate. 'Scientists...may have to stir up public opinion,' he wrote in 1971, 'even lobby for laws before legislatures.' How else would society 'keep pace' with 'the transition of scientific discovery into technological achievement'?

The award of a Nobel prize to Robert Edwards was welcomed throughout the world as richly deserved and long overdue. Many referred to the battles which he and Steptoe had had to fight in bringing IVF in from the cold and giving it a vital momentum which continues to this day. Yet the congratulations for Edwards were not universal. A Vatican official, Ignazio Carrasco de Paula, President of the Pontifical Academy for Life, was reported in Italy as describing the Nobel prize as 'completely out of order', and that 'without Edwards there would be no market for human eggs; without Edwards there would not be freezers full of embryos waiting to be transferred to a uterus, or, more likely, used for research or left to die'. Such sentiments were echoed by other pro-life groups, one saying in a UK press release that 'IVF has made it possible to search out and destroy disabled embryonic children'.

The latter comments illustrate our own understanding of the ethics of IVF, that most attitudes are largely determined by religion and its associated cultural opinions. The Catholic church has consistently opposed IVF because it removes conception from sexual intercourse and, because the embryo has the moral status of a human being from the moment of conception, has the potential to destroy life. It is this same embryo protection principle which drives the Roman Catholic church's opposition to abortion and certain types of contraception—and why in some countries attitudes towards abortion (as in Ireland where it is illegal) or indeed to IVF (as in Italy's infamous Law 40 of 2004) may be more restrictive than elsewhere. However, even in countries with a less dominating Catholic culture—as in Germany—the freezing of embryos is still perceived as an offence against human life.

Such views, of course, are not always acceptable to the individual, whether Catholic or not, nor consistent with the declared positions of the United Nations or World Health Organization on human rights. It is also clear that assisted reproduction is acceptable to many other Christian religions (because of the 'blessing' of children), as it is to the Jewish and Islamic faiths (providing in the latter it is performed solely within the context of husband and wife).

However, what particularly underlay the criticism of Robert Edwards—albeit in very isolated circumstances—is the extent to which IVF has necessitated the discarding of embryos, either to avoid multiple pregnancy or, if genetically affected, following PGD analysis. Both of these adverse outcomes to conception (spontaneous or assisted) are central to this book and in our mind their avoidance is justified on two grounds: first, for the better welfare of the child by avoiding future harm and illness; and second, for the enhanced autonomy of the parents in helping them have a child who is healthy and not likely to be a psychological (and physical) burden on them. As we have repeatedly made clear, our support for PGD does not imply a support for eugenics; PGD is about individuals and their own choices, eugenics is not. In our view PGD and the avoidance of multiple pregnancies are central to 'designing' babies today, and are justified as applied technologies in the welfare of the children born and the parents' freedom of choice once they are in full possession of the facts.

It is for these latter reasons that we could not condone the transfer of embryos known to be affected by a disability present in the parents, such as deafness. Both the welfare of the child and his/her future autonomy would be undermined by a request to transfer embryos whose genetic composition was known to be consistent with disability. This would apply if, for example, a couple suffering from deafness requested the transfer of similarly affected embryos, or if in the preimplantation diagnosis of sex-linked disease (such as some muscular dystrophies) only embryos of the susceptible gender were available.

The possibility that only affected embryos are available after PGD should certainly be discussed with the parents during pretreatment PGD counselling, so that everyone is clear about options and policy. However, it remains our view that the risk of harm to the child (and parents) is a serious contraindication to the transfer of any affected or susceptible embryos. Thus, in our counselling for PGD it is explained that, if no non-affected embryos are available, the only options are to try a new cycle of treatment or reconsider reproductive plans (donor eggs or sperm, or adoption, perhaps). However, such decisions are less categorically made when the PGD is for the detection of genetic mutations whose consequences are less clearly defined, such as the BRCA1 and 2 genes associated with breast cancer. In such cases the disease is serious but of later onset, the penetrance of the mutation is less predictable, and successful treatments for the disease are available. In these cases attitudes towards the availability of only affected embryos require more latitude, and more individual 'case-sensitive' evaluation.

While safe IVF does imply the discarding of embryos, the ethical questions which this raises are somewhat resolved by timing. The Vatican argues that life begins at conception, from which point the zygote and embryo deserve appropriate recognition and respect. And we too recognise that preimplantation embryos deserve respect because of their potential for life—but embryos at their early cleavage stage, or even at the blastocyst stage, are still at the very rudimentary stages of development. It is scientifically implausible that the union of two cells (the oocyte and sperm as a zygote) can be designated as the beginning of life—or indeed that any single developmental moment can be so called. Even fertilisation itself takes place over a matter of hours, not moments, and the zygote does not become complete until it has implanted in the endometrium. Monozygotic twins, for example, which are formed when a single zygote divides, may be separated in their 'creation' by several days but share a common genomic footprint. Such twins demonstrate that the zygote does indeed represent the *potential* for life, in that two lives may be derived from it.

The UK government, which in its original Human Fertilisation and Embryology Act allowed embryo research for up to 14 days, and later accepted emergency contraception if used within two weeks of conception, has taken an 'embryological' position in that the embryo cannot become an individual being until the third week of pregnancy, when monozygotic twinning is no longer possible. Such views have been challenged, and other arguments (neurological, for example) proposed, but so far science has been unable to provide any conclusive evidence of when life begins, and thus whether the discarding of embryos in IVF and PGD is morally acceptable or not.

What is clear to us is that all reproductive technologies have their benefits and their drawbacks, and the challenge is to judge the good and the bad, to distinguish the value from the non-value. Ultimately, such judgements become inherently subjective, and in certain circumstances difficult to make according to strict scientific criteria. In the case of PGD or single embryo transfer the arguments are largely based on the welfare of the child and parents, with likely outcomes to conception validated by scientific data. In our view the welfare of the child (whether conceived spontaneously or by IVF) is better served by the avoidance of an inherited disease; PGD makes that possible, while spontaneous conception does not. Some have argued that such a view discriminates against inherited disorders, suggesting that children born with Down's syndrome, for example, are less valued, less loved, than children born without problems. This is not our argument. It is simply that PGD allows the choice between the transfer of affected or unaffected embryos, and that choice is a matter of judgement for the parents and the doctor dependent on the welfare of the child.

The doctor (or counsellor) has a role to play in guiding the parents through these moral arguments, respecting their autonomy and presenting the scientific facts (about safety or success, for example). And sometimes, as we saw in chapter 2, the patients' wish may not always overlap with the doctor's answer. Even in a routine case, we would not be willing, for example, to transfer three or four embryos in an IVF procedure to a woman with good prospects of success; the risk of twins is too great.

That, however, presents a 'routine' ethical question. What is more likely to make headlines and divide opinion—and prompt our professional organisations to issue hasty position statements—are those cutting-edge procedures which have not been tested by time and public acceptability. The introduction of IVF and PGD, ICSI, HLA tissue typing, assisted reproduction for lesbians, posthumous IVF, the treatment of HIV positive men and women, and egg donation in postmenopausal women have all been accompanied by controversy and debate as the public tries to 'keep pace' (as Edwards predicted) with the progress of technology. And so far history tells us that the limits to that progress are more defined by public acceptability than by the technology itself.

Cloning in the age of Dolly

In February 1997 *Nature* carried a report from scientists at the Roslin Institute in Edinburgh that a live lamb had been born from an embryo created not by fertilisation but from the nucleus of an adult cell. During 1996 the Roslin scientists had transferred the nuclei removed from several types of sheep cells into unfertilised sheep oocytes from which their own genetic material had been removed. These renucleated oocytes were then activated (with electric current) to cause cell fusion and cleavage. They were then cultured and transferred to a surrogate mother ewe. Just one of these transferred oocytes implanted (from 277 fused cells prepared), and was carried to term by the surrogate. The resulting lamb—known through-out the world as Dolly—thus carried the exact genetic material of the adult ewe who provided the cell (from her udder), not of the sheep who provided the egg.

What everyone saw once the story was released was a bright, well-grown lamb whose life had been derived not from fertilisa-tion but from the genetic material of a single adult cell. To this extent, Dolly was a clone of the six-year-old ewe who provided the udder cell whose nucleus was transferred and activated in the host enucleated oocyte. Dolly carried all the chromosomes from that donor udder cell and none of the chromosomes from the host/recipient oocyte. Thus, to the public at large it was now apparent that adult cells do have within their struc-ture the genetic material necessary for the production of living organisms. Despite the Roslin's protests that 'nuclear transfer' was not cloning, in the public's perception the age of the clone had dawned. Even *Nature* joined in, headlining its cover with 'A flock of clones'.

For all of us working in reproductive science and medicine, the birth of Dolly would have both direct and indirect lasting consequences. First among them was that now the ethics of reproduction by any assisted technology had become a matter of enormous public debate. The public, of course, was not expected or asked to understand the techniques of nuclear transfer, nor the mysteries of genetics, but it did have a crude understanding that life was now possible without fertilisation and that mammals could be replicated from a single adult cell. What made the ethical impact of Dolly that much more intense was also that science and medicine—like everything else in public life—were now subject to so much more public scrutiny. Accountability was now an item on the agenda of all news organisations, controversy a matter not just for experts but now for the public at large.

In reproduction, the birth of Dolly was important because it demonstrated that life could be created from an adult (somatic) cell. Dolly disproved the belief that only the genes in a fertilised egg are sufficiently totipotent and appropriately activated to generate a new living organism, that adult cells, because so many have become inactivated (or activated) over time, are simply too specialised to create life. The Edinburgh researchers who created Dolly said their objective was to find ways of producing livestock carrying specific genes, which ultimately might guarantee the presence of target proteins in milk. Today, such objectives remain the focus of 'therapeutic cloning', but the technique, now referred to as somatic cell nuclear transfer (SCNT), is most prominently applied in the search for a viable source of embryonic stem cells. And few issues have raised such controversial debate as the use of these technologies to create embryos specifically for stem cell research.

Many public organisations—such as the American Association for the Advancement of Science or World Health Organization—have made pronouncements on human cloning and recognised a distinction between 'reproductive' and 'therapeutic' purposes; it is in the latter that the embryo's inner mass is collected and cultured for the derivation of embryonic stem cells, and these may have therapeutic regenerative applications. There is universal consensus that reproductive cloning in humans (the creation of one human from the adult cell of another) is not acceptable, not least because the technique's safety is totally unproven. UNESCO, the Council of Europe, WHO (resolutions WHA50.37 (1997) and WHA51.10 (1998)), the World Medical Association, the European Union and the European Parliament—as well as numerous individual countries—have all adopted measures banning reproductive cloning.

However, there is far less consensus on therapeutic cloning, nor a common position on its ethical acceptability. Indeed, even the debate itself is presented on a variety of unequal non-overlapping levels, with a 'moral' objection to the creation of stem cells set against the 'practical' health benefits which such research might one day achieve. The central issue, therefore, about human cloning is not whether it should be outlawed, but whether therapeutic cloning should be included in the ban. At the reproduction level (after all, SCNT uses a technique perfected in ICSI and embryo biopsy) ESHRE has in place a 'moratorium' on reproductive cloning, but has accepted the potential of therapeutic cloning as an instrument of regenerative medicine. Similarly, the American Association for the Advancement of Science has described reproductive cloning as 'unconscionable', but 'supports stem cell research, including the use of nuclear transplantation techniques in order to realize the enormous potential health benefits this technology offers'.

Despite such endorsements, there remain many opponents, and at the heart of the objection—as in all embryo research—lies the ethical difficulty of creating embryos whose purpose is not human life but only the derivation of stem cells. As George W. Bush said in supporting a block on federal funding for stem cell research, 'We shouldn't as a society grow life to destroy it'—and therein lies the core of the ethical controversy. When the American company Advanced Cell Technology announced in 2001 that it had cloned a single six-cell human embryo for the derivation of stem cells, Bush told reporters that it was 'morally wrong', an opinion echoed by the Vatican and the European Commission. 'Not everything scientifically possible and technologically feasible is necessarily desirable or admissible,' said the European Research Commissioner at the time, faintly echoing the warning of Robert Edwards 20 years earlier that society is not always equipped to 'keep pace' with the arguments which science throws in its path.

We share the misgivings of ESHRE and others that there are far too many unknowns (both known and unknown) about reproductive cloning to even consider it as a practicality.

And right now we can think of no acceptable indications for cloning which other technologies—such as PGD—could not otherwise achieve. The unacceptable indications—to replicate individuals, even if in fulfilment of some misguided reproductive autonomy—are violations of human dignity inasmuch as people are reduced to little more than a collection of genes.

The creation of stem cells, however, from human embryos (whether from cloning technology or as surplus to IVF) remains a controversy which science has so far been unable to resolve. This is why so much work in stem cell research today is in the search for alternative sources: a non-embryonic source would remove the controversy, if not resolve it. And why too there was so much acclaim in 2007 when two independent research groups (James Thomson in Wisconsin, US, and Shinya Yamanaka in Kyoto, Japan) announced that human stem cells had been induced to pluripotency from adult (non-pluripotent and fully differentiated) cells. The adult fibroblast cells had been 'transfected' with a small number of genes (just four) which had slowly induced pluripotency as otherwise found in embryonic stem cells. The transfection of these induced stem cells (iPSC) had been achieved by using viral vectors as carriers of the genes into the genome of the cell, but this raised fears for future therapeutic safety—that these induced stem cells might themselves be susceptible to disease. Since then new systems for the delivery of genes into the fibroblast cells have been reported, which appear to induce cells with comparable pluripotency to embryonic stem cells but without any viral transfection.

In Brussels at the VUB we too have been working on the der-ivation of stem cells without the need for destroying embryos. It is work which we think worthwhile. Stem cells can be kept in culture in an undifferentiated state for many years, but can later be differentiated into any cell type in the human body. And it's this versatility which has made them so indispensable to fundamental and applied research: to basic embryology and developmental biology, in vitro drug development and testing, and of course regenerative medicine. Already important prog-ress has been reported from various clinical trials.

In 2008 our group in Brussels showed for the first time that a single blastomere biopsied from a four-cell embryo is able to produce efficiently an embryonic stem cell line with all the characteristics of a line derived from a blastocyst. This allowed us to develop a new strategy for the treatment of children with diseases formerly treatable by bone marrow transplanta-tion. Embryos, obtained through IVF from their parents, are biopsied and the biopsied cell used to derive a stem cell line. This line, if shown to be compatible with the tissue type of the child, would produce the differentiated cells for transplantation as regenerative stem cell therapy, without risk of rejection. We already know from PGD that the loss of one cell from an early stage embryo does not affect its capacity to implant in the uterus. Thus, by intervention at this earlier stage and without destroying the embryo, many of the ethical concerns raised by the conventional derivation of embryonic stem cells would be removed.

Meanwhile, hopes of resolving the ethical difficulties of using human embryos to derive stem cell lines seem firmly pinned on the derivation of cells whose pluripotency has been induced by somatic cell reprogramming. And because these induced cells share many of the characteristics of embryonic stem cells (in their pluripotency) they have been credited with the potential to revolutionise regenerative medicine. Their potential clinical benefit lies in their genetic similarity to the individual from whom they were derived, and thus that immune rejection of transplanted cells is unlikely. Doubts persist over the safety of these iPSC, because of the routes by which they were reprogrammed, but reports of non-integrating reprogramming methods have generated some optimism that they may well be safe for clinical application. Recent studies have also suggested that the characteristics of iPSC and truly embryonic stem cells, while sharing a pluripotent ability to differentiate into various human cell types, may not be totally comparable in their molecular development. Thus, while iPSC remain an enticing source of embryo-free stem cells, research with embryonic cells cannot yet be abandoned, and they remain the gold standard.

Pregnancy in older women

Nowhere have the boundaries which define acceptability in reproductive medicine been more challenged than in the fertility treatment of older women. And nowhere have the technologies of reproductive medicine been more acutely illustrated than in the achievement of pregnancy in an older woman whose natural reproductive function has ceased with the menopause. For many, individuals and societies alike, even the concept is unthinkable: an unacceptable event which defies the laws of nature. But it is age more than the loss of reproductive function which prompts the controversy; when the same fertility treatment is applied in younger women with a premature menopause, there is no such public outcry, even if the treatment still defies the laws of natural reproduction.

In June 2008 a woman of 70 living in India claimed the dubious distinction of being the world's oldest mother. She was given fertility treatment to have the child she craved, a son in a family of girls—although in the event male and female twins of very low birth weight were delivered by Caesarean section at 34 weeks. The same year another report from India suggested that a second woman of 70 had also given birth, though 18 months later her health was said to be rapidly deteriorating. This too was the fate of Spanish mother Maria del Carmen Bousada de Lara, who in 2006 and at the age of 66 had become the world's oldest mother following the birth of twin boys; she died three years later of undisclosed causes. In defending her right to have children, she had argued that, even if she did die, her children were part of an extended family and would always be cared for. She had been treated at an IVF clinic in Los Angeles where, according to reports, she had lied about her age.

Age in assisted reproduction as in natural conception is a highly emotive subject. As a guide through this moral maze, there are few rules: the science is inexact and opinions divergent. But what is becoming clearer, as average maternal age increases and more and more couples elect to postpone the birth of their children, the 'ethical' debate is rapidly becoming a matter of greater public and practical expedience. Expedience, as reflected by individual circumstances and the realistic chance of success, is also the route taken by many ethicists considering the issue, who accept that a threshold of 'minimal effectiveness' is appropriate for the evaluation of any treatment within a healthcare system.

Conversely, as we suggested with reference to funding for IVF, ethicists might also ask if it is morally acceptable to provide treatment only to those with a good chance of success? Based on a simple law of fairness, should not everyone have at least one chance (however small) of having a child—and thus at least the chance of treatment? Again, there are no simple answers, and none which science clearly puts forward. In our view, therefore, these difficult ethical issues are best resolved practically and individually, on the basis of treatment outcome, maternal safety (in pregnancy and childbirth), the interests of the child, the disposition of clinics to treat their older patients, and the regulatory framework.

Based on these fundamental considerations, and notwithstanding the restraints of our own clinics' policies, the refusal to treat older postmenopausal women is justified on the grounds of pregnancy complications, and the interests of the child as assessed within a context of parental illness and an overextended generation gap. In the Netherlands legislation restricts treatment to women who are 45 or younger, while in Belgium fresh embryo transfer can only be performed in women of 45 or younger, and frozen embryo transfer at 47 or younger. These cut-off ages are based on medical, pragmatic and legal considerations. Complication rates in any pregnancy over the age of 40 are significantly increased. The clinics which to our knowledge do treat postmenopausal women do so within a framework of intense individual counselling, the general health of the patient, and the involvement of the family doctor.

At the public level the acceptability of postmenopausal pregnancy is more dependent on a gut reaction which evaluates parental age alongside the age of the child: the suitability of 70-year-old parents to a teenage child, the possibility of parental death or illness, and the ability of the mother to deliver the baby in safety. Yet it is also a fact that the parameters of these reflexive reactions are rapidly changing. Even within a single generation, female life expectancy has increased substantially and is now edging beyond 85 years in most developed countries; forecasts are that many baby girls born today will live to see their hundredth birthday. And as female life expectancy has extended, so have female expectations. Careers, relationships, travel are still the ambitions of many postmenopausal women, and still within their reach. But babies?

The technology to produce a pregnancy in a woman beyond the menopause is no longer a reproductive challenge—providing that she's happy to accept a donor egg and the certainty that the baby will carry none of her own genes. Data from the US—where payment to egg donors is allowed—show that (fresh or frozen) donor eggs were used in 12 per cent of all assisted reproduction treatments performed in 2007. There, the use of donor eggs increases sharply after the age of 39; and, among women older than 44, about 75 per cent of all fertility treatments now use donor eggs.

Conventional IVF is hopelessly unsuccessful in these women, and prohibitively expensive. For example, an evaluation of 1,101 fresh, non-donor IVF cycles performed in Australia in 2002–04 in women aged 45 years or over (as reported at ESHRE's annual meeting in 2007) calculated an overall success rate of less than 1 per cent but, even more alarmingly, at a huge cost per live birth achieved. Reflecting a treatment failure rate of more than 99 per cent, these costs were 29 times the overall average cost of a live birth following fresh IVF treatment in Australia. For the record, 21 women achieved a clinical pregnancy, but 71 per cent of these were lost prematurely. There were just six live births, a rate of 0.5 per 100 started cycles. Similar results have been reported from other countries. In the UK, for example, a study of women having a first cycle of IVF at a clinic in Scotland found the cost of a live birth after IVF rising significantly at the age of 40 because of lower success rates; a mean cost per birth more than 2.5 times higher than that for a woman aged 35–39 years.

Egg donation, however, can achieve far higher success rates in these women. Those same US data for 2007 also show that 52 per cent of transfers using donor eggs in women aged 24 resulted in a live birth, 54 per cent in women at age 45, and a remarkable 51 per cent over the age 48. The figures indeed confirm that the age of the recipient has no effect on implantation potential. Even more remarkably, a 2002 study from California—where egg donation treatment in postmenopausal women seems anyway much less controversial than in Europe—reported a live birth rate of 37 per cent in a series of 77 women all aged over 50. Commenting on the results, the authors (from the University of Southern California) concluded that 'appropriately screened' women can achieve similar pregnancy and delivery rates to younger women, adding: 'There does not appear to be any definitive medical reason for excluding these women from attempting pregnancy on the basis of age alone.'[2]

However, a study of all births in Sweden between 1987 and 2001 did find that peri- and neonatal problems all increased with maternal age, although the incidence was relatively low in all age bands. The findings prompted the investigators to comment: 'Although maternal and fetal outcome is compromised, pregnancies in women aged 40–44 and 45 or older may still be considered relatively safe.'[3] In the California report of egg donation treatment in the 77 over-50s noted above, there was an above-average rate of premature delivery and of low birth weight, with almost two in three births delivered by Caesarean section. Mild pre-eclampsia was noted in 25 per cent of patients and severe pre-eclampsia in 10 per cent. All these complications have been noted as higher than usual in egg donation patients. A study of 199 egg donation treatments at the VUB in Brussels in the 1990s found that the most frequent complications of pregnancy were uterine bleeding in the first trimester (in 35 per cent of patients), hypertension (in 33 per cent) and intrauterine growth retardation (in 12 per cent). These complications were more prominent in twin pregnancies, and their incidence increased with patient age. The Caesarean section rate was 64 per cent.

Nevertheless, it is already clear that the demand for fertility treatment will continue to increase in women over 40, especially as a result of trends in delayed pregnancy, and in rarer instances in women beyond the natural menopause. Assisted reproduction in the form of egg donation will be able to produce successful pregnancies in many of these women, and, providing access to treatment or oocytes is not restricted, is unlikely to be the limiting factor to demand. Indeed, egg donation is widely offered and accepted as a treatment for premature menopause. The limit, therefore, will not be the reproductive technology itself, but more likely a 'social age deadline' for childbearing, which, according to a recent study funded by the European Commission, reflects current attitudes towards reproductive behaviour.[4]

The French demographer Henri Leridon proposed in 2008 that this same social age deadline explains the discrepancy between potential and actual childbearing after the age of 40. Indeed, Leridon proposed that most women in their 40s have not yet reached the end of their biological reproductive potential, and that only 55 per cent of women even at the age of 45 are definitively sterile. But the reasons they are not having the babies they are biologically capable of having are because of a perception that certain behaviours are only appropriate within a certain time limit. These perceptions have been found different from one country to the next, and different with respect to men and women. But they do exist, and are driven in part by the health risks to mother and child we have already described.

This European Commission study questioned more than 20,000 people in 25 European countries and the vast majority of them (96 per cent) cited a mean maternal age deadline of 42 years. This is a 'mean' figure, so there were answers much higher and much lower, but what it does say is that the majority of Europeans simply find it unacceptable on aggregate for women over the age of 42 to be having babies. However, it's worth noting that the number who felt that women should not be having children after age 40 was higher among older than younger respondents, suggesting that these social age deadlines reflect the current social context and will extend in time. The study also found a direct correlation between a later country-specific age deadline and the availability of fertility treatments. We assume that in an environment like California, for example, with fertility technologies easily accessible and donor eggs in plentiful supply, the social age deadline for childbearing will be much later than in countries with no or limited egg donation programmes available.

Posthumous reproduction

Invariably, the ethical dilemmas of assisted reproduction come down to a decision of whether to treat or not to treat. Many ethicists have based their grounds for denying treatment on its futility and the likelihood of a very poor outcome; others pay greater respect to the autonomy of the patient and her right to treatment. But not all the debates in assisted reproduction can be resolved on the ethical cornerstones of beneficence, autonomy and justice. There are other arguments too which demand an individual assessment and the recognition of a 'moral' contraindication to treatment. The welfare of the child, for example, may be detrimentally affected by parental history: alcohol abuse, a criminal record, even psychiatric problems may all be reasons to withhold treatment.

In such circumstances the clinician or counsellor is in a very difficult position. There may be no obvious reason— scientific or regulatory—to deny treatment, but individual counselling may suggest a reason not to go ahead. These are subjective decisions, and inevitably open to question. In Brussels we see this especially in the fertility treatment of single women, which in our opinion still presents one of the most complex ethical situations which fertility clinics face. Around half the single women we see at the VUB for an initial consultation are refused treatment, not because treating singles is not allowed, but because we are not reassured during counselling of the best interests of the child. There are many patients who can't accept our opinion, and who will certainly look elsewhere, but we feel that judgements in such circumstances are better made than not even considered, even though there are no objective criteria or scientific validation to underpin our conclusions. But in these difficult circumstances an understanding of the patient's motives and expectations seems reasonable. It's also a challenge for fertility clinics that in these same circumstances we can only base our opinions on what we know, on what we're told. Some of the headline scandals in reproductive medicine have arisen when patients were less than honest with their doctors.

253

However, if the treatment of single women wanting a baby presents complex ethical questions, they are even more so when the patient is single because of the death of a partner. Posthumous treatment has produced some of the most difficult questions in reproductive medicine, some of which have proved beyond the scope of moral judgements and have only been resolved by the courts. Such cases are rare, but the majority are found in the context of fertility preservation in advance of cancer treatment. In men the sperm will have been produced and stored before death; in women eggs may have been retrieved and stored, or fertilised and stored as embryos. Many jurisdictions have few regulatory problems with treatments which use these stored gametes in the surviving partner provided that the pregnancy meets local regulatory requirements (with respect to consent, parental rights and inheritance) and is the continuation of what is already judged as a pregnancy and parenthood project.

In such cases the previously recorded written consent of the deceased is required for posthumous treatment in Belgium, the Netherlands, Spain and the UK, and some states of the US. In the UK the consent must also name the designated partner. There are also varied restrictions on the time after death by which treatment must be started or completed—started within six months in Belgium, and completed within one year in Spain, two years in the Netherlands and three years in Belgium. However, in all these circumstances the gamete retrieval and storage (and even fertilisation) take place before the death of the partner, with pregnancy and birth taking place after the death.

More ethically complex are the very rare circumstances in which the sperm is retrieved after the death of the male partner. Because sperm quality diminishes with time, the sperm cells must be retrieved as soon as possible after death. Again, a requirement for written consent seems the best way to avoid any future confusions over a posthumous pregnancy. This may seem highly unlikely in the context of accidental death (or vegetative state), but even consent forms for conventional IVF and embryo storage usually raise the possibility of posthumous options. It is our view that in the absence of written consent no action to obtain gametes posthumously should be undertaken.

Even more rare—and even more ethically demanding— are cases in which female gametes have been stored (either as oocytes or embryos) and, following the death of the female partner, require a surrogate for pregnancy and birth, or implantation in new partner. Past written consent is necessary for such cases to go ahead, and surrogacy arrangements will usually require some sort of legal agreement.

However, common to all these circumstances is our concern for the welfare of the child and the psychological condition of the surviving partner, and that should be considered in addition to the requirements of consent and local regulation. The scientific literature is remarkably sparse, so there is little guidance from the evidence of studies. The treatment decision will thus rest on an assessment of the future child's development and well-being in a one-parent family which lost a partner before the request for treatment. From the few studies available, this seems unlikely to be harmful to the child. Indeed, an ESHRE report on the ethics of posthumous reproduction concluded: 'Because the surviving partner will have a positive view of the deceased, it is likely that the story of the child's conception will present the child as a much desired gift issuing from a loving relationship. Contrary to the use of an anonymous donor, the child will know his or her genetic origins and will be given a positive image of the deceased parent.' We have greater concerns about the child's welfare when a surrogate is used following maternal death or when insemination is requested when the male partner is fatally ill. The absence of the mother may be more important in the child's psychological development than the absence of the father, but there are very few studies to guide our judgements.

Indeed, it is this very absence of objective data which makes posthumous reproduction so controversial. But it is a treatment which is technologically possible and which works. Whether it is applied or not will depend on local regulatory requirements, which in themselves are nationally heterogeneous, and a case-by-case assessment of the circumstances. Not all countries accept posthumous treatment, and those that do impose tight requirements (or looser guidelines) over consent and time-frames. It is our view that written consent should have been given by the deceased partner before the use of stored gametes or embryos, and that consent should have been obtained at the time of storage or before the start of the treatment cycle. Thorough counselling and assessment of the surviving partner during the decision-making period is essential.

The technological limits

The headlines which used to accompany cases of posthumous reproduction seem now to have subsided, as legal requirements, consent agreements and guidelines have established more common criteria. Of course, the law cannot always be such a moral arbiter, but in many societies it does reflect the prevailing attitudes of religion or culture. Any fertility treatment which, for example, involved a third party—and would therefore be performed outside the parameter of marriage—would be unacceptable and illegal in a jurisdiction dominated by the Muslim faith.

Yet not all laws, even if reflecting the dominant faith, are necessarily and morally right. Indeed, in 2004 we described as 'unethical' the legislation introduced by the Italian parliament which banned embryo freezing but compelled clinics to fertilise and transfer a maximum of three oocytes. The result was that most clinics were forced to transfer their maximum three embryos, and thereby impose on their patients an unfair risk of multiple pregnancy. It was only through legal challenges in 2009 that this requirement to transfer all embryos produced in a treatment cycle (thereby avoiding 'selection') was found 'unconstitutional', even though the law had been introduced with the backing of the Vatican and following a public referendum. Embryo freezing is now allowed in Italy, but only in the interests of patient health, either physical or psychological. Our misgivings about the fairness of the Italian legislation have since been underlined by data from

ESHRE showing that in 2007 almost 3 per cent of all IVF/ICSI deliveries in Italy were triplet, when in countries such as Belgium, Denmark, Slovenia, Sweden and the UK, the rate was 0.3 per cent or below.

Similarly, in 2009 the European Court of Human Rights in Strasbourg upheld the complaint of two Austrian couples that Austria's embryo protection laws, which banned third-party oocyte and sperm donation, were discriminatory and in violation of the couples' human rights (under two articles of the European Convention on Human Rights). The case had involved two couples seeking fertility treatment in Austria, one of whom required IVF with donor sperm and the other male and female gamete donation. In its judgement, the Court said that the 'wish for a child' is protected by the European Convention, and that its fulfilment through fertility treatment should not be prevented by 'unjustified discriminations'. 'Moral considerations', the Court added, or concerns about social acceptability, 'are not in themselves sufficient reasons for a complete ban on a specific artificial procreation technique such as ova donation'.

The decision created a storm of protest among pro-life organisations, one of which warned that 'If this decision is upheld [on appeal], the flood gates will open for the recognition of a protected right for same sex couples to access artificial procreation with egg or sperm donors exactly like a couple composed of a man and a woman'.

Legislation usually removes (though rarely resolves) the ethical decisions which clinics would otherwise have to make, but it's clear that even the law is not always morally infallible.

It is also clear, in the numbers of patients who seek treatment in foreign countries, that many will pass judgement on their home country's regulations by simply side-stepping the domestic restriction and going abroad. One of the arguments in favour of cross-border reproductive care is that it homogenises the ethical patchwork of national jurisdictions and gives patients an equal opportunity of treatment, even if at a price.

But the law cannot intervene in those areas where we see the biggest challenges facing IVF in the years to come, and that's in dealing with the demands of an ever ageing fertility population and its attendant decline in oocyte and embryo quality. Regulation or policy might put a cap on treatment in women over 45, but it cannot legislate against the ever growing number of women who put their family plans on hold and, at the age of 38 or so, find themselves unable to conceive. A very recent study performed in the fertility clinic of Maastricht University in the Netherlands has found that the average age of patients attending a first consultation there increased from 27.7 years in 1985 to 31.4 years in 2008, a rise of almost four years in just two decades. Moreover, the proportion of women of 35 years or older using the clinic's services almost quadrupled; in 1985 it was 7.9 per cent and in 2008 had increased to 31.2 per cent. Maastricht is not alone. These are trends seen in fertility clinics throughout the world and the challenge in facing them is one of reproductive technology, not ethics.

The success and efficiency of IVF continue to increase slowly but surely year on year, and to some extent are able to absorb the demands of an ageing patient population. But it is an unfortunate fact of reproductive life—as the databases of the

HFEA, CDC or ESHRE clearly illustrate—that IVF success rates in women over 40 are and will remain very low. There is little that we can do to improve oocyte or embryo quality, which is why technology is now concentrating on embryo selection and implantation, in trying to identify those factors consistent with pregnancy and thereby make our treatments even more efficient.

And it is in this ever growing patient population that the boundaries will be set by technology, by how successful we can make assisted reproduction, and not simply by its public acceptability. But for those women whose ovarian reserve is already depleted, the basic technologies of IVF and ICSI will be inadequate for the creation of a biological child; their only viable fertility treatment will require egg donation. And in the technique of egg donation our technologies, policies and guidelines have not yet met the challenge of supplying sufficient donor eggs without the inducement of money.

These are the indelible circumstances of IVF today. It is tempting, as many clinics find, simply to transfer more embryos in order to compensate for the decline in fertility which comes with age. In some cases such an approach seems quite reasonable, as the latest guidelines of the ASRM suggest. But even the transfer of three embryos in a 45-year-old woman, although compliant with the guidelines, may still present a risk of multiple pregnancy which we consider unacceptable. Today's challenge, therefore, is to strike a balance between a good prognosis and the risk of harm. And in that equation, we believe, the overriding consideration is to avoid the risk of harm.

Steps towards the risk-free IVF clinic

Yesterday	**Tomorrow**
Down-regulation with a GnRH agonist	Immediate suppression with a GnRH antagonist
hCG for final egg maturation	GnRH agonist for final egg maturation
Fresh embryo transfer	Freeze all, transfer later in a natural or artificial cycle
Double/triple embryo transfer	Single embryo transfer after thawing
OHSS = 5%	OHSS = 0%
Multiple pregnancy = 20%	Multiple pregnancy = 0%

Technology has served IVF well in providing the means to avoid harm. Today, the concept of a risk-free IVF clinic is not far from our grasp—and the risks of ovarian hyperstimulation and of multiple pregnancy can each be virtually eliminated. By adopting an approach which substitutes a GnRH antagonist for an agonist, which uses a GnRH agonist for triggering oocyte maturation, which follows a policy of single and not double embryo transfer, and which freezes all embryos derived from a first treatment cycle and transfers later, we have the framework for safe IVF. The cost may be a modest fall in birth rates, but the benefit will be the certainty that this is a close as we can get to risk-free treatment, and this is what we consider the responsible application of science in the 'design' of babies in our fertility programmes.

POSTSCRIPT
Paul Devroey

It is easy to challenge our interest in the investigation and treatment of infertile men and women. Indeed, one may even ask if reproductive medicine has any role at all in an overpopulated world of 7 billion people. Notwithstanding the fact that so many people suffer from malnutrition and a lack of the very basics of life, the treatment of infertile couples trying to circumvent primary or secondary sterility is not easily justified.

However, the interpretation of overpopulation on the one hand and an individual need on the other requires a balanced judgement, and it is my view that infertile couples should be able to overcome their childlessness through the medical therapies which lead to pregnancy. It is also important to remember that an infertile couple is very different from an individual with disease. In general, infertile couples do not physically suffer from their condition, even though their infertility causes a great deal of emotional and psychological distress. It is remarkable to me that when we treat an infertile couple there are always two people involved. This scenario is different from almost all other aspects of medicine, where the doctor-patient relationship is invariably one-to-one.

In many clinics today the male and female partner of the infertile couple are investigated by two different doctors. So it was a privilege for me to be trained in gynaecology and in androl-ogy at the University Campus of the VUB in Brussels, where my professor at that time, Robert Schoysman, was integrating the treatment of both male and female infertility. It gave me the huge advantage of being able to perform male and female investiga-tions, as well as microsurgical operations in both. This combined training is not widespread today, which I think is a great mistake.

It is also remarkable how much our progress in these natural sciences is so closely linked to the motivation and passion of individuals. Such is the case with my own collaboration with André Van Stierteghen, which led to the development of ICSI and TESE. Another exceptional individual is Professor Robert Edwards, who was really at the beginning of every important development. It is fair to say that without him much of the progress we have charted in this book would not have been possible. Personally, I have no doubt that his vision inspired a completely new era in reproductive medicine, and also laid the firm basis for the training and clinical developments which we all enjoy today. Robert Edwards was also the founder of the European Society of Human Reproduction and Embryology (ESHRE), and within ESHRE I had the honour to work with him for many years. It was he who persuaded me that the eth-ics of reproduction could never be ignored and that the science of our work, its creativity and innovation, must be transparent and well defined. It is true that in the clinic we know the result of treatment very very soon—it's always a yes or no to preg-nancy—but what lies behind that treatment must be recorded for our peers as an effort of mutual fertilisation between scien-tists and clinicians. This is a step-by-step process of publication, transparency and communication at scientific meetings and in our medical journals. Progress in science is not difficult provided we ask the right questions at the right moment.

Not only is reproductive science unusual because it deals with two people, it also deals with two gametes, the oocyte and sperm, and in the fact that the most interested person in the equation is the child to be. Certainly, many players are involved in this process and of course, the embryo itself is an extremely symbolic object for society in general and for the philosophical and cultural movements in those societies. It is of great interest to me to see how many different religions have different opinions about the status of the embryo. Some faiths, such as Roman Catholicism, cannot accept IVF at all and reject any related treatment. Others, such as the Muslim, Buddhist and Hindu faiths, accept IVF. These different responses have huge ramifications for the implementation of different reproductive therapies throughout the world. The restrictive laws introduced in Italy in 2004 were inspired by the Vatican, the embryo protection laws of Germany inspired by cultural historical events. Such examples illustrate how reproductive science has such an unusual place in our society today.

As a scientist in reproductive medicine I have always approached my work from the basis of science: peer-reviewed publications; well-established journals; the uptake and reproducibility of an innovation throughout the world. It's a position which sees science as the driver for research in a non-profit organisation. I have had the advantage of working all my life at a fixed salary in a state hospital. The focus was only on science and the implementation of that science in daily practice.

Throughout that time there have been many research projects in Brussels which have been fascinating and compelling, but they were all based on very simple questions, with yes or no answers, and an effort to improve what we already had. And when I reflect on these innovations, it is remarkable how quickly the activities of today become tomorrow's past experience. Great innovations such as oocyte donation and the cryopreservation of human embryos in Australia, preimplantation genetic diagnosis in Britain, the use of GnRH agonists and antagonists, were all hailed at the time as remarkable breakthroughs, but are now a part of our everyday repertoire.

In Brussels our greatest innovation was intracytoplasmic sperm injection, or ICSI, yet behind its development lay the very basic fact that traditional IVF was (and is) only possible as a treatment for female infertility if the sperm of the male partner is normal. ICSI opened up a whole new field of treatment for almost all infertile men, and, through the simple introduction of a single sperm cell into a single oocyte, revolutionised practice throughout the world. Suddenly, men who formerly had no chance of becoming a father of their own genetic child, now had that opportunity made possible. Many thousands of children have been born since we first described ICSI in 1992 and many millions will be born in the future.

I vividly remember working with André Van Steirteghem, Gianpiero Palermo and Sherman Silber, and thinking—as did many scientists at the time—that a sperm needed to pass the epididymis to reach full maturation. But I still remember very clearly how, when we did a biopsy of the testis in one case of obstructive azoospermia, the sperm moved after a few moments beneath the microscope. These same sperm cells were injected and provided the world's first pregnancy with testicular sperm, a technique we called testicular sperm extraction (TESE). The report was published in *Fertility and Sterility* in 1996 and from then on it became very clear that even men with complete obstructive azoospermia could become the fathers of their own children. Today, TESE is widely used in combination with ICSI. In addition, in our treatment of men with obstructive azoospermia who were also carriers of cystic fibrosis, it became possible not only to overcome the obstruction but also to detect the disease in the embryo. From this moment on, preimplantation genetic diagnosis (PGD) in combination with ICSI and TESE could be performed without the risk of disease in the child.

Without doubt, the goal of our reproductive technology treatments is to make such treatments safe. And it's fascinating to see just how much safer these treatments have become over time. We still have two big enemies: multiple pregnancies and the occurrence of ovarian hyperstimulation syndrome (OHSS) but it's gratifying to realise that with new therapies and strategies both these outcomes can be eliminated. The risk of multiple pregnancies can be almost totally removed by replacing fewer and fewer embryos; replacing just one will lead to a huge reduction. Ovarian hyperstimulation syndrome, which occurs in approximately 2 per cent of all stimulated cycles, is a life-threatening condition which can be eliminated by the use of GnRH antagonists (and an agonist only to trigger oocyte maturation). It could now be perceived as almost unethical not to implement such a strategy. How can one accept even a single case of OHSS knowing that a medical strategy is available to avoid it completely?

One of the most important developments of recent years is the improvement in cryopreservation. The vitrification of human oocytes has not just made possible high implantation rates after fertilisation and better embryo development, but also now opens up the era of oocyte banking. This will have a big impact in preserving the fertility of cancer patients. In addition, the vitrification of human embryos will have an important role to play in the introduction of safer IVF. The best environment in which to transfer an embryo is the natural or artificial, not stimulated, cycle; a policy to freeze all fresh embryos and transfer one at a time in subsequent cycles will improve outcome, and safety.

Which leads me to the concept of an OHSS-free clinic in a singleton delivery environment. Thus, the future of IVF will probably be in a combination of three different steps: first, the ovarian stimulation protocol optimised for egg collection and the avoidance of OHSS; second, the cryopreservation of human oocytes and embryos; and third, their replacement one by one in an artificial cycle. This is a concept also applicable to egg donors, who need never again run the risk of OHSS.

These are not novel technologies, but they are innovative because they allow the adaptation of our therapies to the patient's indications and safety. For me, this is the way forward for IVF, the continued improvement of pregnancy outcomes in the safe environment of singleton pregnancies and an OHSS-free clinic.

Paul Devroey
Brussels, February 2011

Bart Fauser

My earliest recollection of IVF is seeing the birth of the Netherlands' first IVF baby on the Dutch national news. The 'gynaecologist' involved was not even a gynaecologist, but a second-year resident from Rotterdam! The scientist involved, Gerard Zeilmaker, had been unable to find an established gynaecologist willing to work with him and develop this new IVF technique. At the time I too was a resident in obstetrics and gynaecology—but my subsequent subspecialty training focused mainly on reproductive endocrinology—full of ideas for PhD projects aiming to understand the normal function of the male and females gonads. I was thrilled by such fascinating developments, which motivated me to pursue an academic career.

My focus in basic and applied clinical science over the last three decades has moved from the neuroendocrine regulation of testis function (my PhD), to ovarian physiology (extensive laboratory studies in the US), to the understanding and treatment of women with ovarian dysfunction (especially polycystic ovarian syndrome, PCOS), to the development of mild strategies for IVF, and finally to new concepts involving the (usually ignored) gender differences in general medicine, particularly in cardiovascular disease and ageing.

The way infertility was treated when I began my career seems far in the past now. Despite initial resistance and disbelief, IVF has proved to be a major development, as reflected in the long overdue Nobel prize awarded to Robert Edwards in 2010. Today, IVF has trivialised most diagnostic procedures and outweighed almost all other fertility treatments.

Throughout the 1980s and early 1990s distinct improvements in IVF were made by fine-tuning ovarian stimulation regimes, the introduction of embryo cryopreservation, and oocyte donation. These basic advances were accompanied by genuine breakthroughs in assisted reproductive technology with the development of PGD, ICSI and the use of testicular sperm, and more recently egg freezing. These have all given new hope to couples who previously had almost a zero chance of pregnancy. Not surprisingly, all these breakthrough have been widely publicised in the popular press.

Although IVF is now one of the most documented (and regulated) medical procedures, it has clearly drifted away from the paradigms of conventional medicine. In the later 1990s, when I seriously considered how I could best contribute to the field of IVF, it struck me then that the era of big breakthroughs was over. IVF had found its place as a mainstream fertility treatment, and an ever increasing number of big (usually private) centres around the world were (and still are) providing progressive, high quality services. IVF has taken an empirical approach to infertility, aiming to offset the decline in natural, age-related female fertility with a slow but steady improvement in results.

However, IVF's intense and self-contained focus on short-term 'success' (usually defined as pregnancy rate per cycle) leaves little room for reflection on how assisted reproduction can advance further, nor on any real—and by definition more subtle—implications for the future health of the patient and her baby. But this concern is now my real motivation—and what keeps me awake in reflection. Like Paul Devroey, I have worked in an academic environment (with a fixed salary) all my professional life, and an advancement in the broad scope of reproductive medicine has always been my major inspiration. With this aim in mind I believe we have a duty to make IVF less hazardous for the woman and her baby, less complex and more affordable, and we should all make efforts to improve worldwide access to treatment.

Bart Fauser
Utrecht, February 2011

EIGHT CHAPTERS IN A NUTSHELL

One may ask why we felt the need to write this book, and why we did it together. We both have a long-standing scientific relationship: in addition to almost two decades of joint reflection, we have organised combined research meetings since 1997 for both our departments, and numerous national and international scientific conferences. Bart Fauser was appointed a visiting professor at the VUB in Brussels in 1995, Paul Devroey similarly appointed in Utrecht in 2008.

So far, our joint collaboration has resulted in 36 scientific papers in eight different international journals (including *Human Reproduction, Fertility and Sterility, Human Reproduction Update, The Lancet, Reproductive Biomedicine OnLine,* and *Journal of Clinical Endocrinology and Metabolism*). Our joint research covers areas such as drug development for ovarian stimulation, the implications of ovarian stimulation for egg and embryo quality, luteal phase endocrinology and endometrial receptivity.

We are very aware that the international networks in reproduction mainly involve scientists and fertility specialists, despite the fact that developments have considerable implications for society. Yet the public and policymakers are often insufficiently informed, let alone actively involved in shaping the future of fertility treatments.

It was this paradox which prompted the idea of this book, to explain the broader implications of fertility treatment today: the technologies themselves, demographic trends, the effects of age, the ethical controversies, preimplantation genetic diagnosis, multiple pregnancies, and oocyte and embryo cryopreservation. They all arose from our long discussions in Rotterdam, London, Ostend and Rome over a near two-year period in which the book took shape.

TAKE HOME MESSAGES

Fertility trends

- Humans are poor reproducers (pregnancy rate per cycle is low), a remarkable observation in the context of the densely populated world we live in.

- There is a clear trend in developed societies for women to have fewer children. More and more women (especially those who are higher educated) decide not to have children at all.

- There is an ongoing trend for women to postpone childbearing. Increasing female age is associated with a falling chance of pregnancy. As a result, more women are in need of fertility treatment.

IVF: where are we now?

- IVF has revolutionised infertility treatment.

- The era of major improvements in IVF is over, and research now focuses on fine-tuning.

- Current challenges involve:
  a) improving the safety of treatment (reducing ovarian hyperstimulation syndrome and multiple pregnancy);
  b) reducing the burden of treatment for the woman and her child;
  c) improving global access to treatment (by making IVF more affordable and broadening insurance coverage).

- A redefinition of success in terms of a healthy baby born.

IVF: the new horizons

- Improved cryopreservation techniques for both oocytes and embryos will encourage completely different IVF strategies, and will allow women to preserve their fertility (both for medical and other reasons).

- New molecular technologies may soon become available, improving patient selection for treatment, embryo selection, endometrial receptivity, and outcomes.

PHYSIOLOGY: THE WOMAN

Embryo The product of fertilisation in which the fertilised egg has the full complement of DNA from both parents. At the first stage of cell division, the fertilised egg is known as a **zygote**; as cell division increases after two or three days, the embryo is at its cleavage stage; and at day five or six of cell division it reaches the **blastocyst** stage.

Endometrium The lining of the uterus, into which the embryo will implant to form a pregnancy. In the early menstrual cycle the endometrium thickens in readiness for implantation.

Fallopian tubes Two tubes which connect each ovary with the uterus. After ovulation, the egg passes down one of the Fallopian tubes, where fertilisation takes place.

Follicle (antral) A small ovarian follicle at an early stage of its development, following maturation through the primordial and pre-antral phases. Antral follicles are visible on ultrasound and are counted as a measure of ovarian reserve or of response to ovarian stimulation in IVF.

Follicle (ovarian) Collections of cells within the ovary from which an egg (oocyte, ovum) will mature, resulting in ovulation from the leading follicle. Several hundreds of follicles will mature each month, but in most cases only a single leading follicle will mature all the way to ovulation. Women are born with a fixed number of follicles (around 2 million) and this follicle pool is continuously depleted throughout life. These follicles can never be replaced. The supply is exhausted at the menopause.

Menstrual cycle A regular menstrual cycle lasts 28 days and begins with the first day of menstrual bleeding. Ovulation usually occurs at mid-cycle; up to that point, the cycle is described as **follicular** (or proliferative, when follicles continue to mature and the lining of the uterus thickens in preparation for pregnancy), and after ovulation is described as **luteal** (or secretory, when hormone levels subside ready for menstruation). A condition without menstrual bleeding is known as **amenorrhea**, often caused by polycystic ovarian syndrome.

Oocyte The female **gamete** (or germ cell) produced in the ovary, which matures to become a fully developed egg cell. Oocyte as a term is often used synonymously with egg (as in oocyte donation/egg donation).

Ovary The female reproductive organ from which an egg (oocyte, ovum) is released each month. Women have two ovaries, at each side of the pelvic region, and oocyte release (**ovulation**) occurs once during each 28-day cycle, from either the left or right ovary.

PHYSIOLOGY: THE MAN

Epididymis A long coiled tube within the scrotum in which sperm cells mature. The epididymis is one source for surgically extracted sperm cells in men whose infertility is caused by a tubal obstruction.

Sperm The male gamete, found and delivered in semen. Around 100 million sperm cells are usually contained in the semen of a healthy single ejaculate, and throughout the course of a reproductive lifetime men will produce many billions of sperm in the testis. The head of the sperm cell contains enzymes and undergoes a reactive process which enables the sperm cell to penetrate an egg for fertilisation. The condition in which no sperm is present in the semen is known as **azoospermia**.

Testis The two testes produce sperm, as well as the male sex hormone, testosterone.

HORMONES

Androgens Generic term for any hormones which develop and maintain male characteristics (such as testosterone).

Estrogen The female sex hormone, secreted by the ovary. Estrogen helps development of the female characteristics, regulates the menstrual cycle and prepares the uterus for pregnancy.

Follicle stimulating hormone FSH, a reproductive hormone produced by the pituitary gland in men and women. Tests of natural FSH provide some indication of fertility (high levels are associated with the menopause, for example). In the ovary follicles mature under the influence of FSH; in men, FSH is involved in the production of sperm. In assisted reproduction manufactured gonadotrophins such as FSH are used to stimulate the ovaries to produce several eggs.

Luteinising hormone LH, a reproductive hormone produced by the pituitary gland in men and women, but important in women at mid-cycle for ovulation. In assisted reproduction the natural surge in LH before ovulation must be suppressed (with GnRH agonists or antagonists) to ensure the maturation of several follicles. LH is the prime driver of sperm production in men.

Progesterone A hormone produced by the ovary in the second half of the cycle. Along with estrogen, progesterone helps maintain the menstrual cycle and prepare the uterus for pregnancy (or menstruation).

GENETICS

Chromosome The fine filament-like structure found in every cell which contains inherited genes. Normally, chromosomes are paired in each cell, a total of 23 pairs in the human, one half inherited from the father, one half from the mother. One of these pairs, chromosomes X and Y, determine gender; females have two X chromosomes, males one X and one Y. The mother always contributes one X chromosome, the father either an X or Y (so the father determines the child's gender). Chromosomal disorders—such as Down's syndrome—occur when these paired arrangements are abnormal (either too few or too many copies).

Epigenetics The concept of how an individual's genetic code and phenotype can be affected be factors other than inherited genes.

Genes Genes are made up from a long strand of DNA, which is copied and inherited over many generations. DNA is composed of repeating units which carry the necessary information to make proteins; genes are copied each time a cell divides. Children inherit one copy of a gene from the father, and one copy from the mother. Genetic problems happen when mutations occur in genes; in some cases, such as cystic fibrosis, the mutation may be inherited, or it may arise over time.

Genotype Characteristics which are due entirely to genetic inheritance; phenotype describes a visible manifestation dependent on both inherited genes and physical traits.

FERTILITY TREATMENT

Clomiphene citrate The most common drug used in fertility treatment, especially for inducing ovulation in women with anovulatory polycystic ovary syndrome, and as an empirical treatment in women with regular cycles to generate multiple eggs.

Cryopreservation Storage in deep-freeze conditions, at temperatures of almost −200°C; developed for the preservation of sperm or surplus embryos in IVF. Most clinics today are adopting the new rapid freezing technique of **vitrification**, for both embryos and oocytes.

Egg donation A fertility treatment designed for women unable to produce their own eggs (because of depleted ovarian reserve or **primary ovarian insufficiency**, POI). Eggs are provided by an anonymous or known donor, and fertilised with sperm from the patient's own partner.

Gonadotrophins Man-made versions of natural reproductive hormones. FSH (follicle stimulating hormone) and hMG (human menopausal gonadotrophin, which contains similar amounts of FSH and LH) are both used to stimulate growth of several follicles within the ovary for IVF and ICSI.

GnRH agonists/antagonists Man-made hormones designed to suppress the activity of the body's own reproductive hormones released from the pituitary gland. GnRH agonists (administered as a nasal spray or injection) take around two weeks to achieve their suppressive effect; GnRH antagonists work with immediate effect.

Intracytoplasmic sperm injection ICSI, a 'micromanipulation' technique in which a single sperm cell is injected into an egg to achieve fertilisation; although originally developed for the treatment of male infertility, ICSI is becoming the world's favoured fertilisation technique for all forms of infertility.

Intrauterine insemination IUI, a technique of 'artificial' insemination in which a sample of semen (from the partner or anonymous donor) is injected through a catheter into the uterus to help natural fertilisation.

In vitro fertilisation IVF, the original technique of assisted reproduction in which eggs (in follicular fluid) are aspirated from the ovary and mixed with sperm cells to achieve fertilisation in the laboratory; after two to five days, one or two embryos are transferred to the uterus for pregnancy.

Polycystic ovarian syndrome PCOS, a condition associated with infertility defined by any two of three symptoms: multiple cysts on the ovary; high levels of male (androgen) hormones; or absent/irregular periods.

Preimplantation genetic diagnosis PGD, a technique which screens an individual couple's embryos (created by ICSI) for a specific single gene or chromosomal defect. The aim is to avoid an inherited disease.

Preimplantation genetic screening PGS, a technique which screens multiple IVF embryos by genetic technologies such as **FISH** (fluorescence in situ hybridisation) or **CGH** (comparative genomic hybridisation). The aim is to improve embryo selection for IVF, to increase the chance of pregnancy and make single embryo transfer more efficient.

Premature ovarian failure POF, also known as primary ovarian insufficiency, which is evident as a menopause before the age of 40; this may occur either spontaneously or as a result of cancer treatment.

ORGANISATIONS

ASRM The American Society for Reproductive Medicine, responsible for the most binding treatment guidelines in the US.

ESHRE The European Society of Human Reproduction and Embryology, Europe's foremost society in reproductive science and medicine responsible for many position papers, clinical guidelines and training.

HFEA The Human Fertilisation and Embryology Authority, the UK's regulatory agency, responsible for ensuring that the requirements of fertility legislation are met by registered clinics (which include the collection of data from all IVF and ICSI cycles performed).

NOTES

Chapter 1

1. Shin M, Besser LM, Kucik JE, et al, 'Prevalence of Down syndrome among children and adolescents in 10 regions of the United States' *Pediatrics* 2009; 124: 1565–71.
2. Palermo G, Joris H, Devroey P, Van Steirteghem AC, 'Pregnancies after intracytoplasmic injection of single spermatozoon into an oocyte' *Lancet* 1992; 340: 17–18.
3. Steptoe PC, Edwards RG, 'Birth after the reimplantation of a human embryo' *Lancet* 1978; 2: 366.

Chapter 2

1. Shenfield F, de Mouzon J, Pennings G, et al, 'Cross border reproductive care in six European countries' *Human Reproduction* 2010; 25: 1361–8.
2. Domar AD, Zuttermeister PC, Friedman R, 'The psychological impact of infertility: a comparison with patients with other medical conditions' *Journal of Psychosomatic Obstetrics and Gynaecology* 1993; 14 Suppl: 45–52.
3. Chen T-H, Chang S-P, Tsai C-F, Juang K-D, 'Prevalence of depressive and anxiety disorders in an assisted reproductive technique clinic' *Human Reproduction* 2004; 19: 2313–18.
4. Verberg MF, Eijkemans MJ, Heijnen EM, et al, 'Why do couples drop-out from IVF treatment? A prospective cohort study' *Human Reproduction* 2008; 23: 2050–5.
5. Van Peperstraten AM, Hermens RPMG, Nelen WLDM, et al, 'Perceived barriers to elective single embryo transfer among IVF professionals: a national survey' *Human Reproduction* 2008; 23: 2718–23.

6. Pinborg A, Loft A, Rasmussen S, et al, 'Neonatal outcome in a Danish national cohort of 3438 IVF/ICSI and 10 362 non-IVF/ICSI twins born between 1995 and 2000' *Human Reproduction* 2004; 2: 435–41.

7. Hvidtjorn D, Grove J, Schendel D, et al, 'Multiplicity and early gestational age contribute to an increased risk of cerebral palsy from assisted conception: a population-based cohort study' *Human Reproduction* 2010; 25: 2115–23.

8. Veleva Z, Karinen P, Tomás C, et al, 'Elective single embryo transfer with cryopreservation improves the outcome and diminishes the costs of IVF/ICSI' *Human Reproduction* 2009; 24: 1632–9.

9. Lintsen AME, Pasker-de Jong PCM, de Boer EJ, et al, 'Effects of subfertility cause, smoking and body weight on the success rate of IVF' *Human Reproduction* 2005; 20: 1867–75.

Chapter 3

1. Gnoth C, Frank-Herrmann P, Freundl G, et al, 'Time to pregnancy: results of the German prospective study and impact on the management of infertility' *Human Reproduction* 2003; 18: 1959–66.

2. 'A prospective multicentre trial of the ovulation method of natural family planning. III. Characteristics of the menstrual cycle and of the fertile phase' *Fertility and Sterility* 1983; 40: 773–8.

3. Templeton A, Morris JK, Parslow W, 'Factors that affect outcome of in-vitro fertilisation treatment' *Lancet* 1996; 348: 1402–06.

4. Baird DT, Collins J, Egozcue J, et al, 'Fertility and ageing' *Human Reproduction Update* 2005; 11: 261–76.

5. Munné S, Sultan KM, Weier HU, et al, 'Assessment of numeric abnormalities of X, Y, 18, and 16 chromosomes in preimplantation human embryos before transfer' *American Journal of Obstetrics & Gynecology* 1995; 172: 1191–9.

6. Belloc S, Cohen-Bacrie P, Benkhalifa M, et al, 'Effect of maternal and paternal age on pregnancy and miscarriage rates after intrauterine insemination' *Reproductive Medicine Online* 2008; 17: 392–7.

7. Carlsen E, Giwercman A, Keiding N, Skakkebaek NE, 'Evidence for decreasing quality of semen during the past 50 years' *British Medical Journal* 1992; 305: 609–13.

8. Auger J, Kunstmann JM, Czyglik F, Jouannet P, 'Decline in semen quality among fertile men in Paris during the past 20 years' *New England Journal of Medicine* 1995; 332: 281–5.

Chapter 4

1. Gardner DK, Schoolcraft WB, Wagley L, et al, 'A prospective randomized trial of blastocyst culture and transfer in in-vitro fertilization' *Human Reproduction* 1998; 13: 3434–40.
2. Khalaf Y, El-Toukhy T, Coomarasamy A, et al, 'Selective single blastocyst transfer reduces the multiple pregnancy rate and increases pregnancy rates: a pre- and post-intervention study' *British Journal of Obstetrics and Gynaecology* 2008; 115: 385–90.
3. Rotterdam ESHRE/ASRM-Sponsored PCOS consensus workshop group, 'Revised 2003 consensus on diagnostic criteria and long-term health risks related to polycystic ovary syndrome (PCOS)' *Human Reproduction* 2004; 19: 41–7.
4. Marcoux S, Maheux R, Berube S, 'Laparoscopic surgery in infertile women with minimal or mild endometriosis' *New England Journal of Medicine* 1997; 337: 217–22.
5. Copperman AB, DeCherney AH, 'Turn, turn, turn' *Fertility and Sterility* 2006; 85: 12–13.

Chapter 5

1. Rijnders PM, Jansen CA, 'The predictive value of day 3 embryo morphology regarding blastocyst formation, pregnancy and implantation rate after day 5 transfer following in-vitro fertilization or intracytoplasmic sperm injection' *Human Reproduction* 1998; 13: 2869–73.
2. Guerif F, Le Gouge A, Giraudeau B, et al, 'Limited value of morphological assessment at days 1 and 2 to predict blastocyst development potential: A prospective study based on 4042 embryos' *Human Reproduction* 2007; 22: 1973–81.
3. Baxter Bendus AE, Mayer JF, Shipley SK, et al, 'Interobserver and intra-observer variation in day 3 embryo grading' *Fertility and Sterility* 2006; 86: 1608–15.
4. Mastenbroek S, Twisk M, van Echten-Arends J, et al, 'In vitro fertilization with preimplantation genetic screening' *New England Journal of Medicine* 2007; 357: 9–17.
5. Schoolcraft WB, Fragouli E, Stevens J, et al, 'Clinical application of comprehensive chromosomal screening at the blastocyst stage' *Fertility and Sterility* 2010; 94: 1700–06.

6. Jones GM, Cram DS, Song B, et al, 'Novel strategy with potential to identify developmentally competent IVF blastocysts' *Human Reproduction* 2008; 23: 1748–59.

7. Market-Velker BA, Fernandes AD, Mann MR, 'Side-by-Side Comparison of Five Commercial Media Systems in a Mouse Model: Suboptimal In Vitro Culture Interferes with Imprint Maintenance' *Biology of Reproduction* 11 August 2010; 83: 938–950.

Chapter 6

1. Handyside AH, Kontogianni EH, Hardy K, Winston RM, 'Pregnancies from biopsied human preimplantation embryos sexed by Y-specific DNA amplification' *Nature* 1990; 344: 768–70.

2. Handyside AH, Lesko JG, Tarín JJ, et al, 'Birth of a normal girl after in vitro fertilization and preimplantation diagnostic testing for cystic fibrosis' *New England Journal of Medicine* 1992; 327: 905–09.

3. Dahl E, Gupta RS, Beutel M, et al, 'Preconception sex selection demand and preferences in the United States' *Fertility and Sterility* 2006; 85: 468–73.

4. Gleicher N, Barad DH, 'The choice of gender: is elective gender selection, indeed, sexist?' *Human Reproduction* 2007; 22: 3038–41.

5. Puri S, Nachtigall RD, 'The ethics of sex selection: a comparison of the attitudes and experiences of primary care physicians and physician providers of clinical sex selection services' *Fertility and Sterility* 2010; 93: 2107–14.

6. Coleman MP, Quaresma M, Berrino F, et al, 'Cancer survival in five continents: a worldwide population-based study (CONCORD)' *Lancet Oncology* 2008; 9: 730–56.

7. Lee SJ, Schover LR, Partridge AH, et al, 'American Society of Clinical Oncology recommendations on fertility preservation in cancer patients' *Journal of Clinical Oncology* 2006; 24: 2917–31.

8. Forman EJ, Anders CK, Behera MA, 'A nationwide survey of oncologists regarding treatment-related infertility and fertility preservation in female cancer patients' *Fertility and Sterility* 2010; 94: 1652–6.

9. Homburg R, van der Veen F, Silber SJ, 'Oocyte vitrification—Women's emancipation set in stone' *Fertility and Sterility* 2009; 91(4 Suppl): 1319–20.

10. Martinez-Burgos M, Herrero L, Megias D, et al, 'Vitrification versus slow freezing of oocytes: effects on morphologic appearance, meiotic spindle cofiguration, and DNA damage' *Fertility and Sterility* 2011; 95: 374–7.

11. Donnez J, Dolmans MM, Demylle D, et al, 'Livebirth after orthotopic transplantation of cryopreserved ovarian tissue' *Lancet* 2004; 364: 1405–10.
12. Ernst E, Bergholt S, Jorgensen JS, Andersen CY, 'The first woman to give birth to two children following transplantation of frozen/thawed ovarian tissue' *Human Reproduction* 2010; 25: 1280–1.
13. Silber SJ, DeRosa M, Pineda J, et al, 'A series of monozygotic twins discordant for ovarian failure: ovary transplantation (cortical versus microvascular) and cryopreservation' *Human Reproduction* 2008; 23: 1531–7.
14. Golombok S, Badger S, 'Children raised in mother-headed families from infancy: a follow-up of children of lesbian and single heterosexual mothers, at early adulthood' *Human Reproduction* 2010; 25: 150–7.

Chapter 7

1. Hoorens S, Gallo F, Cave JAK, Grant JC, 'Can assisted reproductive technologies help to offset population ageing? An assessment of the demographic and economic impact of ART in Denmark and UK' *Human Reproduction* 2007; 22: 2471–5.
2. Connolly M, Gallo F, Hoorens S, Ledger W, 'Assessing long-run economic benefits attributed to an IVF-conceived singleton based on projected lifetime net tax contributions in the UK' *Human Reproduction* 2009; 24: 626–32.
3. Jungheim ES, Ryan GL, Levens ED, et al, 'Embryo transfer practices in the United States: a survey of clinics registered with the Society for Assisted Reproductive Technology' *Fertility and Sterility* 2010; 94: 1432–6.
4. Flisser E, Scott RT, Copperman AB, 'Patient-friendly IVF: how should it be defined?' *Fertility and Sterility* 2007; 88: 547–9.

Chapter 8

1. Johnson MH, Franklin SB, Cottingham M, Hopwood N, 'Why the Medical Research Council refused Robert Edwards and Patrick Steptoe support for research on human conception in 1971' *Human Reproduction* 2010; 25: 2157–74.
2. Paulson RJ, Boostanfar R, Saadat P, et al, 'Pregnancy in the sixth decade of life: obstetric outcomes in women of advanced reproductive age' *Journal of the American Medical Association* 2002; 288: 2320–3.

3. Jacobsson B, Ladfors L, Milsom I, 'Advanced maternal age and adverse perinatal outcome' *Obstetrics & Gynecology* 2004; 104: 727–33.
4. Billari FC, Goisis A, Liefbroer AC, et al, 'Social age deadlines for the childbearing of women and men' *Human Reproduction* 2011; 26: 616–22.

INDEX